Soil-Steel Bridges

Soil-Steel Bridges

Design and Construction

George Abdel-Sayed
Professor of Civil Engineering
University of Windsor

Baidar Bakht
Principal Research Engineer Ministry of
Transportation of Ontario and
Adjunct Professor of Civil Engineering
University of Toronto

Leslie G. Jaeger
Emeritus Professor of Civil Engineering
and Applied Mathematics
Technical University of Nova Scotia
and Associate Vaughan Engineering
Associates, Halifax

McGraw-Hill, Inc.

New York   San Francisco   Washington, D.C.   Auckland   Bogotá
Caracas   Lisbon   London   Madrid   Mexico City   Milan
Montreal   New Delhi   San Juan   Singapore
Sydney   Tokyo   Toronto

Library of Congress Cataloging-in-Publication Data

Soil-steel bridges : design and construction / edited by George Abdel
-Sayed, Baidar Bakht, Leslie G. Jaeger.
      p.   cm.
   Includes bibliographical references and index.
   ISBN 0-07-003021-9
   1. Bridges, Tubular—Design and construction. 2. Bridges, Iron
and steel—Design and construction. 3. Underground construction.
4. Soil-structure interaction. 5. Sheet-metal corrugated.
I. Abdel-Sayed, George. II. Bakht, Baidar. III. Jaeger, Leslie G.
TG390.S65   1993
624′.4—dc20   93-17143
                                                       CIP

1 2 3 4 5 6 7 8 9 0 DOC/DOC 9 9 8 7 6 5 4 3

ISBN # 0-07-003021-9

*The sponsoring editor for this book was Larry G. Hager and the
production supervisor was Pamela A. Pelton. It was set in Century
Schoolbook by North Market Street Graphics.*

Printed and bound by R. R. Donnelley & Sons Company.

This book is printed on acid-free paper.

For the Canadian Society for Civil Engineering

Contents

Chapter 3. Mechanics of Behavior   43
Baidar Bakht, Aftab A. Mufti and Leslie G. Jaeger

Chapter 4. Structural Design Philosophy   71
George Abdel-Sayed

Chapter 5. Structural Design Procedures   135
George Abdel-Sayed, Baidar Bakht, Ron P. Parish

Chapter 8. Special Features   237
Baidar Bakht, George Abdel-Sayed   000

Chapter 9. Hydraulic Design   273
M. Saeed Choudhary, Ron P. Parish

Chapter 10. Distress, Monitoring, and Repairs   291
Baidar Bakht, John Maheu

Appendix A. Principles of Geotechnical Investigation   337
Cameran Mirza

Appendix B. Flexural Behavior of Bolted Connections   343
Leonid Mikhailovsky, D.J. Laurie Kennedy

Contributors

George Abdel-Sayed
Professor of Civil Engineering
University of Windsor

Baidar Bakht
Principal Research Engineer
Ministry of Transportation of Ontario, and
Adjunct Professor of Civil Engineering
University of Toronto

Bozena B. Budkowska
Associate Professor of Civil Engineering
University of Windsor

M. Saeed Choudhary
Senior Hydrology Engineer
Ministry of Transportation of Ontario

Leslie G. Jaeger
Professor of Civil Engineering
and Applied Mathematics
Technical University of Nova Scotia, and
Associate
Vaughan Engineering Associates., Halifax

D. J. Laurie Kennedy
Professor of Civil Engineering
University of Alberta

Jens H. Madsen
Retired Manager of Engineering
Armtec, Guelph, Ontario

Leonid Mikhailovsky
Senior Structural Engineer
Ministry of Transportation of Ontario

Cameran Mirza
President
Strata Engineering Corporation, Ontario

Aftab A. Mufti
Professor of Civil Engineering and
Director of CAD/CAM Centre of Nova Scotia
Technical University of Nova Scotia

John O'Brien
Regional Bridge Engineer
Alberta Department of Transportation and Utilities

Ron P. Parish
Principal
Bolter Parish Trimble Ltd.
Edmonton, Alberta

Preface

Soil-steel bridges are also known by several other names, e.g., corrugated metal culverts, buried pipe structures, flexible pipes. The use of these structures as short-span bridges is so widespread that there are tens of thousands in existence in North America alone. These structures are also used extensively in Europe.

Despite their widespread use, the design of these structures is often empirical. Many engineers still believe that these structures can be "designed" by selecting sizes from trade handbooks. Good construction procedures, which are essential to the integrity of these structures, are known only to some organizations who have developed them from their own trial-and-error experience. It is now widely recognized that a large number of existing soil-steel bridges are in distress because of general ignorance regarding their design and construction practice.

About 10 years ago the Canadian Society for Civil Engineering (CSCE) formed a technical committee on soil-steel structures under the chairmanship of B. Bakht. The committee consisted of members from North America who are known for their interest in and expertise on different aspects of soil-structure interaction, namely, G. Abdel-Sayed, L. D. Baikie, C. S. Desai, G. Gagnon, F. C. Harvey, L. G. Jaeger, M. G. Katona, R. Krizek, G. A. Leonards, J. Madsen, C. Mirza, A. S. Nowak, J. O'Brien, R. P. Parish, W. A. Porter, R. E. Redshaw, E. T. Selig, and R. S. Standley. The current chairman of the committee is G. Abdel-Sayed.

The purpose of the CSCE technical committee was to provide a forum where state-of-the-art knowledge and information could be exchanged on the specific subject of soil-steel bridges. At the first meeting of the committee, it became obvious that, despite the abundance of technical papers, there was no document available which could be used by an engineer to familiarize himself or herself with all the necessary aspects of soil-steel bridges. Soon after its formation, the CSCE technical committee made an attempt to write a book on the subject, assigning different chapters to different authors. This effort was quickly abandoned, mainly because the resulting material was more a collection of technical papers than a coherent book.

A few years ago some core members of the CSCE technical committee revived the idea of writing the book on soil-steel bridges; they took it upon themselves to write the various chapters in collaboration with experts in specific fields, not all of whom are members of the committee. The result of their efforts is in your hands.

In order to give help rather than confusion to the readers, a special effort has been made to ensure that there is continuity of information and consistency of terminology between the various chapters.

This book has brought together most of the relevant knowledge available in the world regarding the design and construction of soil-steel bridges. By using it, an engineer can not only design a safe and economical soil-steel bridge, but can also avoid the pitfalls that lead to eventual distress, and sometimes failure, in these structures. Specific reference is made in the book to the relevant design codes of the United States and Canada. It is hoped that the book will be found useful by both the practicing engineer and the student.

The book contains useful information (Chap. 10) dealing with the repair and rehabilitation procedures for soil-steel bridges in distress. Using the repair techniques given in this chapter, large sums of money can be saved by avoiding the expensive replacements that would otherwise be necessary.

George Abdel-Sayed
Baidar Bakht
Leslie G. Jaeger

Acknowledgments

The editors owe debts of thanks to a large number of people who have materially assisted in the preparation of this volume. Their number is too large for them all to be acknowledged here; however, the editors do wish to thank by name the following persons for the services mentioned:

- Frank Pianca for reviewing Chap. 2
- Karen Conrod and Maria Feehan for typing the bulk of the manuscript
- Santosh Sharma for the typing of Chap. 8
- Nancy Barkley and Anne-Marie Bartlett for typing the earlier drafts of Chaps. 4 and 5.
- Patti Smurthwaite for the typing of Appendix B
- Cindy Lucas for coordinating the work of drawing the figures
- Periya Naidu and Peter Lesner for preparing most of the drawings

Their services were given cheerfully and skillfully.

Finally, the editors, in common with the authors of the various chapters and appendixes, wish to thank their respective employers for the scholarship and research which have led to this book.

Soil-Steel Bridges

Introduction and History

Jens H. Madsen

Retired from Armtec
Guelph, Ontario, Canada

Baidar Bakht

Ministry of Transportation of Ontario
Downsview, Ontario, Canada

1.1 Historical Overview

James H. Watson was a sheet metal worker in Crawfordsville, Indiana. One evening, after spending a day trying to design a storm water sewer in sheet metal, he visited the local drugstore where he watched the druggist wrap a bottle of medicine in a piece of corrugated cardboard in the manner shown in Fig. 1.1. From the corrugated cardboard wrapping, he got the idea of making the storm water sewer with corrugated metal sheets. Watson discussed his idea with his friend E. Stanley Simpson who was the city engineer of Crawfordsville. Simpson liked the idea and together they started to work on the details of the corrugated metal pipe.

The first attempt at making the corrugated metal pipe was not very successful, as the pipe was shaped like the corrugated cardboard wrapping of the medicine bottle, as shown in Fig. 1.2. This pipe, because of having corrugations running along its axis, had little stiffness against radial pressures of the soil. Then Watson and Stanley realized that the pipe should have circumferential corrugations in order to provide the necessary extra stiffness against lateral loads.

In 1886, J. H. Watson patented the corrugated metal pipe under U.S. Patent No. 559,642, which came to be known as the Watson patent. A few days after securing the patent, Watson assigned 50 percent of the rights to the patent to his coinventor, Simpson. The invention of Watson and Simpson led to the development of a very significant and successful

Figure 1.1 A bottle wrapped in corrugated cardboard.

industry which has since been responsible for the tens of thousands of structures throughout the world that are used primarily as drainage conduits, but also as bridges. It is interesting to note that a Russian publication (*Kolokoloff,* Moscow Transport, 1973) states that buried corrugated metal pipes having diameters up to 1070 mm (42 in) were proposed in Russia in 1875. The concept was applied in a railway project in 1887.

Figure 1.2 The first corrugated metal plate pipe looked like this.

While the ideas put forward by Watson and Simpson now enjoy wide acceptance, their efforts met with little success in the early years. A few corrugated metal pipes had been erected in the vicinity of Crawfordsville before Watson's death in 1889, but these structures were made of iron by the puddling process. It is interesting to note that some of these structures were still in service in the 1970s.

Watson, in trying to introduce the new concept, found a strong supporter of it in W. Q. O'Neill, who was a prominent farmer. About three years after Watson's death, O'Neill purchased Watson's interest in the patent from his estate, and a year later he also purchased Simpson's interest in the patent rights.

Having obtained the full rights to the patent, O'Neill, with the help of his friends, established an enterprise for producing commercially corrugated metal pipes for application in road drainage. His enterprise, which led to the establishment of several associated factories in many states, is responsible for the extensive corrugated metal plate pipe industry of today.

Corrugated metal plate pipes can be divided generically into two categories, namely, those that are manufactured in closed pipe shapes and those that are assembled on site from curved corrugated plates. Clearly, the structures in the former category, which are usually referred to as *corrugated steel pipes,* are normally of smaller diameter than those of the latter category, which are usually referred to as *structural plate corrugated steel pipes.*

In the early days, corrugation profiles were not standardized. Up to the early 1930s, the corrugated metal plate pipes were manufactured from 610-mm- (2 ft) wide plates which were curved and riveted into 610-mm- (2 ft) long pipe elements; these elements were subsequently riveted together into lengths that were suitable for transportation. Field splicing of the shop-manufactured lengths of pipes was achieved by a variety of coupling arrangements.

In 1935 a machine was introduced for constructing pipes of continuous lengths from rolls of plain metal sheets. These pipes had helical corrugations and lock seams, and had diameters of up to 530 mm (21 in). Machines that could produce pipes of larger diameter were introduced in the 1950s.

The structural plate pipe which was assembled from corrugated curved plates on the site was introduced in 1931 for a pipe with a diameter of 2.67 m (105 in). Since that time, the structural plate pipe, which is the principal subject of this book, has been used consistently for short-span bridge applications, both culverts and grade separation structures.

The story of the corrugations being oriented wrongly in the first attempt, which is symbolically shown in Fig. 1.2, speaks eloquently of the fact that the two inventors relied solely upon "gut feeling" and empirical knowledge in eventually arriving at the final, very efficient, solution. It is interesting to note that most of the developments in structures made out of corrugated plates during the subsequent nine decades were similarly inspired by "gut feeling" and empirical knowledge.

1.2   Terminology

Despite the one hundred years' existence of structures made of corrugated metal plates, their terminology is not familiar to all structural engineers, mainly because of lack of standardization. These structures are referred to by a variety of names such as *flexible pipes, buried pipes, buried structures, structural plate culverts,* etc. Standard terminology pertaining to the structures under consideration, as well as their definitions, are introduced subsequently. The same terminology is used consistently throughout the book.

When the Ontario Highway Bridge Design Code was introduced in 1979, it was decided to refer to the structures under consideration by the generic name of *soil-steel structures,* defined as being those bridges which are comprised of structural steel plates and engineered soil, designed and constructed to induce a beneficial interaction of the two materials. In this book, the Ontario term is used, except that the word *structures* is replaced by *bridges,* so that the structures under consideration are referred to as *soil-steel bridges.* Two examples of typical soil-steel bridges are shown in Fig. 1.3*a,b,* showing a culvert and a grade separation bridge. It may be recalled that the term *culvert* refers to a structure which is used for conveying water through its opening.

It is noted that the Ontario Code defines a *bridge* as being any structure having a span of 3 m (9.80 ft) or more that forms part of a highway over, or under, which the highway passes. For the purpose of this book, the restriction of having spans greater than or equal to 3 m (9.80 ft) is not necessary. However, the book is restricted to those bridges in which the plate corrugations are annular and which are assembled in the field from individual corrugated plates. In particular, structures with helical corrugations are not considered.

The term *conduit* is often interpreted to mean the metallic shell of a structure; in this book it is used only as a general term for the bridge opening in a soil-steel bridge.

With reference to conventional bridge superstructures, the *longitudinal direction* refers to the direction of traffic flow on the bridge. In contrast with this usual definition, the *longitudinal direction* in a soil-steel bridge is the same as the direction of the conduit axis, illustrated in Fig. 1.4. The *transverse section,* or the *cross section,* lies in a plane

(a)

(b)

Figure 1.3 Examples of soil-steel bridges: (*a*) a culvert; (*b*) a grade separation bridge.

perpendicular to the longitudinal direction. It can be appreciated that when the conduit axis is inclined, as it often is, the cross section of the structure does not lie in a vertical plane.

1.2.1   Corrugation profile

Corrugated plates commonly used in soil-steel bridges have annular corrugations of a standard profile which is shown in Fig. 1.5. As can be seen in this figure, the pitch and depth of this profile are 152 and 51 mm (6 and 2 in), respectively, with the tolerances permitted by the industry being as shown in the figure. The corrugation profile consists of circular segments of 29 mm (1⅛ in) radius joined by straight segments. This corrugation is usually referred to as the 152 × 51 mm (6 × 2 in) corrugation. It is noted that a deeper corrugation profile has been proposed recently.

1.2.2   Terminology relating to the cross section

The terminology relating to the cross section of a soil-steel structure is illustrated in Fig. 1.6, with the definitions and explanations of the various terms being given as follows:

Conduit wall is the metallic shell of the soil-steel bridge, it being noted that the complete shell is also referred to as the *pipe.*

Crown is the highest point on a transverse section of a conduit.

Haunch is the portion of the conduit wall between the spring line and the top of the bedding, or between the spring line and the footing in the case of an arch structure.

Invert is the portion of the conduit wall contained between the haunches.

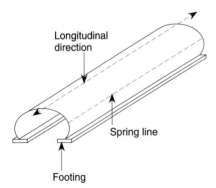

Figure 1.4 Illustration of longitudinal direction.

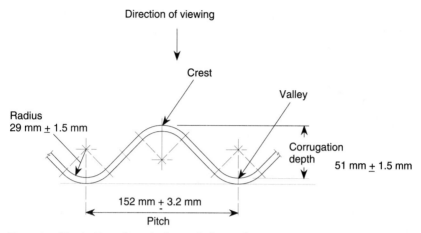

Figure 1.5 Illustration of terminology relating to the corrugation profile.

Invert elevation is the lowest point on a transverse section of a conduit.

Pipe is another term used for the completed conduit wall.

Rise is the maximum vertical clearance inside a conduit at a given transverse section. In the case of structures having a fill inside the pipe to provide a flat riding surface, the rise is measured by ignoring the fill inside the pipe.

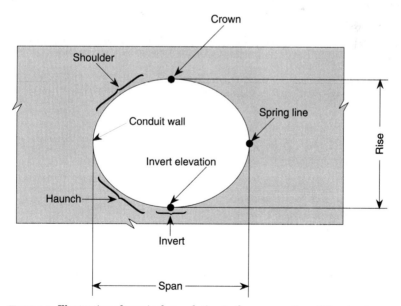

Figure 1.6 Illustration of terminology relating to the cross section of the structure.

Shoulder is the portion of the conduit wall between the crown and the spring line.

Span is the maximum horizontal clearance inside a conduit at a given transverse section.

Spring line is the locus of the horizontal extremities of transverse sections of the conduit.

1.2.3   Terminology relating to soil

The terminology relating to the soil component of the soil-steel bridge is illustrated in Fig. 1.7 and defined as follows:

Backfill is the envelope of engineered soil, excluding the bedding, placed around the conduit in a controlled manner. As discussed in several other places in the book, the backfill must be well-compacted granular soil.

Bedding is the prepared portion of the engineered soil on which the conduit invert is placed. As can be seen in Fig. 1.8, the bedding is contoured to the same shape as the invert. The granular soil of the bedding is kept loose so as to facilitate the soil eventually coming in contact with both the valleys and ridges of the corrugated plate.

Compaction is the process of soil densification, at a specified moisture content, by the application of loads through kneading, tampering, rodding, or vibratory action of mechanical or manual equipment.

Depth of cover is the vertical distance between the profile grade, including the pavement and the crown.

Engineered soil is the selected soil of known properties placed around the conduit in a prescribed manner. Generally, the engi-

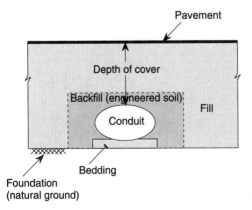

Figure 1.7 Illustration of terminology relating to soil.

Figure 1.8 Photograph showing shaped bedding.

neered soil has to be granular and capable of being compacted to the specified degree.

Fill is the general term used for all the soil placed around the conduit, part of which has to be of a specified quality and compaction while the rest need not be.

Foundation is the soil or rock material underlying the engineered soil.

1.2.4 Structures with closed conduit wall sections

Similar to the earlier corrugated metal pipes, most modern soil-steel bridges have conduit walls of closed sections. The various conduit shapes that are commonly employed are shown in Fig. 1.9, and are individually described as follows:

Round pipe.   As the name implies, the conduit of this structure is circular or nearly circular. The typical cross section of such a structure is shown in Fig. 1.9a. Round pipes of diameters up to about 8 m (26 ft) are in existence.

Horizontally elliptical pipe.   In this structure the conduit is not strictly an ellipse, but is formed out of circular segments. As shown in Fig. 1.9b, the top and bottom segments of the conduit wall have a larger radius of curvature than the two side segments. The conduit has the general appearance of an ellipse with the major axis being

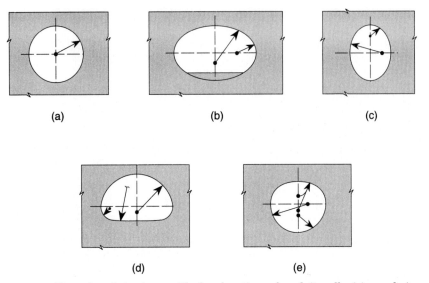

Figure 1.9 Examples of structures with closed sections of conduit walls: (*a*) round pipe; (*b*) horizontally elliptical pipe; (*c*) vertically elliptical pipe; (*d*) pipe-arch; (*e*) inverted pear-shaped pipe.

horizontal. A horizontally elliptical pipe during construction is shown in Fig. 1.10.

Vertically elliptical pipe. Similarly to the horizontally elliptical pipe, this structure is also composed of circular segments with the difference that in this case the side segments have a larger radius of curvature than the top and bottom segments, so that the major axis of the nearly elliptical conduit is vertical. A typical cross section of a vertically elliptical pipe is shown in Fig. 1.9*c*.

Pipe-arch. The typical cross section of a pipe-arch is shown in Fig. 1.9*d*. As can be seen in the figure, the top segment of the conduit wall, which is in the shape of a circular arch, is joined to the invert, which has a relatively large radius of curvature, through haunch segments which are also circular but which have a relatively small radius of curvature. This conduit shape is very popular for drainage applications mainly because of its hydraulic efficiency. A typical pipe-arch structure is shown in Fig. 1.11.

Pear-shaped pipe. This is similar to the vertically elliptical pipe except that the radius of curvature of the top segment of the conduit wall is larger than that of the bottom segment. A typical cross section of this kind of pipe is shown in Fig. 1.9*e*, and a view of such a structure during construction is shown in Fig. 1.12. As can be seen in the latter figure, this particular pipe carries a railway line through it.

Figure 1.10 A horizontally elliptical pipe during construction.

Figure 1.11 A pipe-arch structure.

Figure 1.12 An example of a pear-shaped pipe during construction.

1.2.5   Structures with open conduit
wall sections

For relatively large span and small rise, a soil-steel bridge with closed section of conduit wall becomes uneconomical. In this case the conduit wall section is kept open by supporting the ends of the wall on concrete footings. There can, indeed, be many variations of conduit shapes, but three of the commonly used shapes are shown in Fig. 1.13.

Part arch.   The cross section of a part-arch structure is shown in Fig. 1.13*a*. In this figure it can be seen that the arch is circular and the segment length of the arch is much smaller than that of a semicircular arch.

Half-circular arch.   As the name implies, this structure has a semicircular form. Fig. 1.13*b* shows that for this structure the conduit

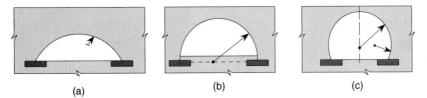

(a)   (b)   (c)

Figure 1.13 Examples of structures with open sections of conduit wall: (*a*) part arch; (*b*) half circular arch; (*c*) reentrant arch.

Figure 1.14 Twin semicircular arch pipes.

wall is vertical where it meets the concrete footing. A structure with twin semicircular arches is shown in Fig. 1.14 during construction.

Reentrant arch. In a reentrant arch, the spring lines are above the junctions of conduit wall and concrete footings. As shown in Fig. 1.13*c*, the conduit wall is composed of three circular segments, with the haunch segments having smaller radius of curvature than that of the top segment.

1.2.6 Conduit walls

As mentioned earlier, the metallic shell of a soil-steel bridge is assembled on the site by bolting together a number of curved corrugated plates. It is shown schematically in Fig. 1.15 that the seams parallel to the conduit axis, or longitudinal direction, are called the *longitudinal seams* and those along the circumference of the conduit, the *circumferential seams*. Usually, the seams are staggered, as can be seen in Fig. 1.16, which shows a pipe during assembly.

Figure 1.17 illustrates the nomenclature for a plate segment according to which the dimension of the segment in the longitudinal direction is called its *length* and the other dimension of the segment when it is straight is termed its *width*. The figure also shows schematically the positions of bolt holes for both the longitudinal and circumferential seams. As can also be seen in the figure, the distance between the centerlines of bolts for circumferential seams is the *net length* of the plate,

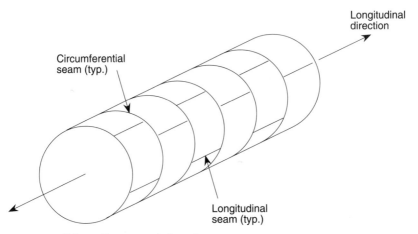

Figure 1.15 Schematic representation of seams.

Figure 1.16 A pipe during assembly.

and, similarly, the distance between the middle of the two rows of bolts, for each of the circumferential seams, is called the *net width* of the plate. The various shop tolerances for the length and width of individual plates are also indicated in Fig. 1.17.

Standard arrangements for bolt holes in individual plates are shown in Fig. 1.18, in which it can be seen that for circumferential seams the bolt holes are always spaced at 244 ± 3 mm (9.6 ± ⅛ in), and the bolt holes for longitudinal seams are always in double rows with a center-to-center spacing of 51 ± 1.6 mm (2 ± ¹⁄₁₆ in). As also shown in the figure, there are three standard arrangements for bolt holes for longitudinal seams. In one arrangement, which is shown in Fig. 1.18a, the bolt holes on successive ridges and valleys of the corrugation alternate between the two rows. In the second arrangement, which is shown in Fig. 1.18b, there is a bolt in every ridge and valley in the row closer to the longitudinal edge of the plate, and in the other row there is a hole either in every valley or on every ridge. In the third arrangement,

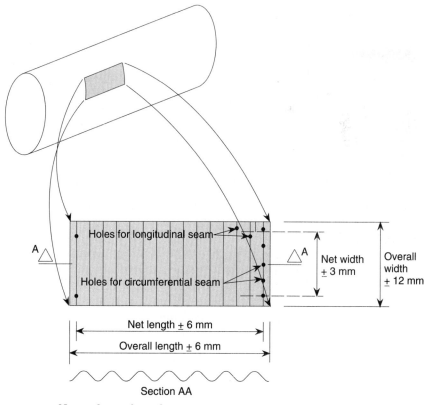

Figure 1.17 Nomenclature for a plate segment.

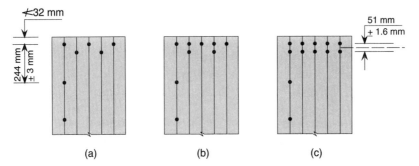

Figure 1.18 Standard bolt hole arrangements: (*a*) arrangement 1; (*b*) arrangement 2; (*c*) arrangement 3.

shown in Fig. 1.18*c,* there is a hole in each valley and on each ridge in each row.

It may be noted that the most popular bolting arrangement is the one shown in Fig. 1.18*a.*

1.2.7 Curvature of conduit walls

All plates are curved into circular segments with the outer 120-mm (4¾-in) lengths being kept straight so as to facilitate the joining of the lapping segments at the longitudinal seams. As shown in Fig. 1.19, while the nominal radius of curvature is with respect to the middepth of the corrugated plate, the curving radius corresponds to the inner extremities of the plate. To avoid the discrepancies in the plate width resulting from the nominal and curving radii, the corrugated plate industry has introduced its own empirical method of calculating the plate width. According to this method (illustrated in Fig. 1.20), with the help of a round pipe, the width of the curved plate at its middepth is assumed to be equal to pi $\times R,$ where R is the curving radius of curvature and pi is an empirical constant which is different from π and whose value is 3.2.

It is interesting to note that, given the depth of corrugations of 51 mm (2 in), the actual value of pi varies with the curving radius of curvature according to the curve shown in Fig. 1.21. The actual value of pi is close to 3.2 only when the curving radius is between about 1.2 and 1.6 m (48 and 64 in). For larger radii of curvature, the value of pi drops down to about 3.15, in which case the use of the empirical method will overestimate the plate width.

Instead of "pi" notation, the industry also uses the N notation, where N is yet another empirical constant which is equal to 3 pi, or 9.6. The plate widths come in only six standard sizes, of which only three are standard fabricator inventory sizes which are available in Canada;

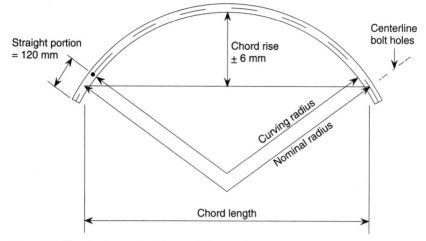

Straight portion = 120 mm

Chord rise ± 6 mm

Centerline bolt holes

Curving radius

Nominal radius

Chord length

Figure 1.19 Terminology pertaining to plate curvature.

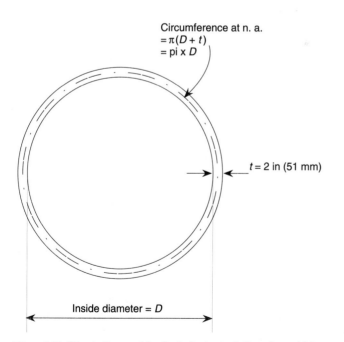

Circumference at n. a.
$= \pi(D + t)$
$= \text{pi} \times D$

$t = 2$ in (51 mm)

Inside diameter $= D$

Figure 1.20 Pi notation used by the industry to define plate widths.

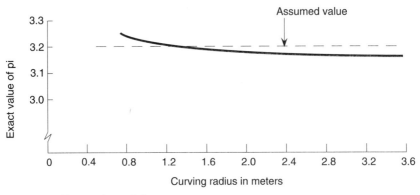

Figure 1.21 Exact values of pi.

these are 15, 18, and 27 pi plates. The standard plate widths are designated as 9, 15, 18, 21, 24, and 27 pi plates, i.e., as 3, 5, 6, 7, 8, and 9N plates. The multipliers to pi and N have units of length in inches. Although the industry in Canada now uses the metric units, the old notation for plate widths is still in use. It is noted that the industry in the United States still uses the United States Customary Unit System. The spacing of bolt holes for circumferential seams is 3 pi or 1N, i.e., 9.6 in or 244 mm, as shown in Fig. 1.18. It may be noted that 24 pi or 8N plate widths are not available in Canada.

For standard plate sizes, the net and overall widths (defined earlier and shown in Fig. 1.17) are listed in Table 1.1. This table also shows the number of bolts for circumferential seams for each of the standard plates.

1.2.8 Other commonly used terms

Commonly used terms pertaining to soil-steel structures, other than those defined earlier in this chapter, are as follows:

Arching. The effect produced by the transfer of vertical pressure between adjoining soil masses which move relative to each other. Positive arching is that which results in the transfer of loads away from the conduit; negative arching produces the opposite effect. Further discussion of the phenomenon of arching is given in Chap. 3.

Beveled end. The termination of the pipe cut at a plane inclined to the horizontal. A beveled end may be square, in which case the pipe is cut symmetrical to the vertical centerline of the conduit; alternatively, the bevel may be skewed, with the pipe on either side of vertical centerline of the conduit being cut nonsymmetrically. It may be

TABLE 1.1   Standard Plate Widths

Nominal width in pi notation	Nominal width in N notation	Net width in inches (millimeters)	Overall width in inches (millimeters)	No. of holes per plate for circumferential seam
9 pi	3N	28.8 (732)	33.6 (853)	4
15 pi	5N	48.0 (1219)	52.8 (1341)	6
18 pi	6N	57.6 (1463)	62.4 (1585)	7
21 pi	7N	67.2 (1707)	71.8 (1824)	8
24 pi*	8N	76.8 (1951)	81.6 (2072)	9
27 pi	9N	86.4 (2195)	91.0 (2314)	10

*Not available in Canada.

noted that the skewed bevel is not generally recommended. A typical square bevel can be seen in Fig. 1.22.

Camber.   A measure of adjustment required in the longitudinal profile of the bedding in order to compensate for postconstruction settlement.

Longitudinal stiffeners.   Longitudinal beams connected to the pipe symmetrically on either shoulder. These stiffeners, which are usually made of reinforced concrete, are also known by the trade name of *thrust beams* or *buttresses.*

Figure 1.22 Pipes with square bevel ends.

Relieving slab.   A reinforced or prestressed concrete slab placed in the fill above the conduit; it is usually provided when the depth of cover has to be kept shallower than would otherwise be permitted.

Transverse stiffeners.   Circumferential stiffeners which are attached to the top portion of the conduit wall before commencing the backfill operation. Transverse stiffeners are usually made out of either curved rolled steel beams, or curved corrugated plates which are placed in a ridge-over-ridge fashion on the pipe. While the bolt holes on the stiffeners themselves are punched in the shop, the corresponding holes on the parent plate are usually made by a flame torch.

1.3   Structural Properties of Conduit Wall

The standard profile of corrugations used for the conduit wall in soil-steel bridges is the 152×51 mm (6×2 in) profile which is shown in Fig. 1.5. In the United States where the United States Customary System (USCS) of units is used by the corrugated plate industry, the plate thicknesses are identified by gauge numbers. Seven standard gauge numbers (1, 3, 5, 7, 8, 10, and 12) are used, with the smaller gauge numbers corresponding to thicker plates. The thicknesses of the plates corresponding to the various gauge numbers are shown in Table 1.2, which also shows the relevant structural properties, these being A, the area of cross section per unit length; I_s, the second moment of cross-sectional area per unit length; and r, the radius of gyration.

With the acceptance of metric units, the corrugated steel plate industry in Canada changed its standards for plate thickness. The new thicknesses are specified in whole millimeters ranging from 3 to 7 mm in steps of 1 mm. The relevant structural properties of 152×51 mm corrugated plates having the new thicknesses are given in Table 1.3.

TABLE 1.2   Structural Properties of 152×51 mm Corrugated Plates with Thicknesses Specified in U.S. Customary System of Units

Gauge no.	12	10	8	7	5	3	1
Nominal plate thickness (uncoated), mm	2.7	3.4	4.2	4.7	5.4	6.2	7.0
A, area of cross section per unit length, mm²/mm	3.294	4.240	5.184	5.798	6.771	7.743	8.719
I_s, second moment of cross-sectional area, mm⁴/mm	990.1	1280.9	1575.9	1769.8	2079.8	2395.2	2717.5
r, radius of gyration, mm	17.32	17.37	17.42	17.48	17.53	17.58	17.65

TABLE 1.3   Structural Properties of 152 × 51 mm Corrugated Plates
with Thicknesses Specified in Metric Units

Nominal plate thickness (uncoated), mm	3.0	4.0	5.0	6.0	7.0
A, area of cross section per unit length, mm^2/mm	3.522	4.828	6.149	7.461	8.712
I_s, second moment of cross-sectional area, mm^4/mm	1057.3	1457.6	1867.1	2278.3	2675.1
r, radius of gyration, mm	17.33	17.38	17.43	17.48	17.52

1.4   Examples of Existing Structures

To give the reader an idea of the kinds of soil-steel bridges that exist in North America, details of a few successful structures are now given. The examples presented are those of structures in only one province of Canada, namely Ontario. The examples, which include structures with relatively large conduits, are expected to be useful in establishing the ranges of conduit sizes that are possible for various categories of soil-steel bridges.

1.4.1   Round pipes

Some details of four soil-steel bridges with round pipe are given in Table 1.4, from which it can be seen that spans of up to 7.6 m can be achieved for this conduit shape by using corrugated plates without any special features such as longitudinal or transverse stiffeners. Details of another structure with round conduit are shown in Fig. 1.23.

1.4.2   Horizontally elliptical pipes

Extensive details are given in Fig. 1.24 for a soil-steel bridge with a horizontally elliptical conduit, constructed in 1978 and having a span and rise of 9.37 and 7.22 m. It can be seen in this figure that the structure is used for grade separation and that its pipe is laid at a steep inclination to the horizontal. The use of both longitudinal and transverse stiffeners in this structure should be noted, along with the fact

TABLE 1.4   Example of Soil-Steel Bridges in Ontario with Round Conduits

No.	Span, m	Depth of cover, m	Conduit wall thickness, mm	Special features	Location	Year of construction
1	7.62	1.4	4.7		Hwy. 21	1973
2	6.10	7.6	4.2		Mun. of York	1977
3	6.10	4.2	5.4		Hwy. 129	1977
4	6.10	4.6	5.4	twin conduit	Lampton County	1966

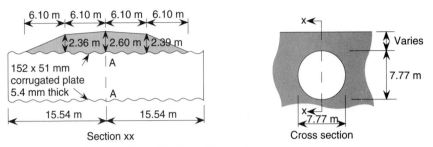

Figure 1.23 Details of a soil-steel bridge with round conduit.

that the plates in the top portion of the pipe are thicker than the plates in the rest of the pipe. It is noted that the terms *granular A* and *granular B* cited in Fig. 1.24 refer to certain well-graded granular soils which are defined in Chap. 4. The example shown in Fig. 1.24 also demonstrates the optimum use of good-quality compacted backfill in a soil-steel bridge.

Examples of a few other structures with horizontally elliptical conduits are given in Table 1.5, from which it can be seen that the conduits for these structures can have spans of up to 10.24 m.

1.4.3 Pipe-arches

Pipe-arches are quite popular for shorter span drainage structures. As can be seen from the details of some examples of pipe-arch structures given in Table 1.6, these structures can have spans of up to about 9 m.

The cross section of a twin pipe-arch structure constructed in 1985 is shown in Fig. 1.25. Of special note in this structure are the concrete inclusions provided in the haunch areas. The pros and cons of these concrete inclusions, and the recommended procedure for casting them, are discussed in Chap. 7. It should be noted that, as will be discussed in Chap. 10, pipe-arches are more prone to distress than other soil-steel bridges mainly because the radius of curvature of the haunch segments of the conduit wall is significantly smaller than the radius of curvature of the invert segment, and also the soil under the haunches is usually very difficult to compact.

1.4.4 Pear-shaped pipes

The cross section of a structure with a pear-shaped pipe is shown in Fig. 1.26. This structure was constructed in 1971 as a railway underpass. A view of the same structure during construction is shown in Fig. 1.12, in which it can be seen that the metallic shell of this structure was first installed with the fill and the railway line inside the pipe to permit the

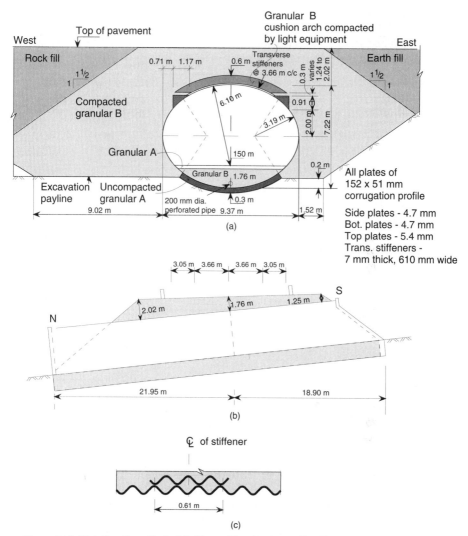

Figure 1.24 Details of a soil-steel bridge with a horizontally elliptical conduit: (*a*) cross section; (*b*) section through conduit axis; (*c*) details of a transverse stiffener.

passage of trains even before starting the backfill operation around the pipe. This structure has a span of 8.53 m, and a rise of 8.84 m.

1.4.5 Arches

When the span of the conduit is required to be considerably larger than the rise, an arch is preferred over pipe-arches and horizontally elliptical conduits. Most of the larger-span soil-steel bridges have open sections of the conduit wall. In the majority of soil-steel bridges with

TABLE 1.5   Example of Soil-Steel Bridges in Ontario with Horizontally Elliptical Conduits

No.	Span, m	Rise, m	Depth of cover, m	Conduit wall thickness, mm	Special features	Location	Year of construction
1	8.76	4.95	1.2	5.4	300-mm-thick relieving slab	Thunder Bay	1973
2	10.24	6.25	2.8	4.7 top segment, 3.4 side and bottom segments	longitudinal & transverse stiffeners	Elgin County	1972
3	8.89	4.37	0.8	4.7 top segment, 3.4 side and bottom segments	longitudinal & transverse stiffeners	County of Essex	1973
4	7.82	4.37	1.4	5.4 top segment, 4.7 side and bottom segments		County of Wellington	1974

TABLE 1.6   Examples of Pipe-Arches

No.	Span, m	Rise, m	Depth of cover, m	Conduit wall thickness, mm	Special features	Location	Year of construction
1	8.99	5.28	3.7	5.4		Prescot City	1966
2	8.31	5.61	7.9	7.0		QEW Southern Ontario	1969
3	7.32	4.70	2.1	4.2	concrete invert	North York	1967
4	6.25	3.91	0.9	4.7		Southern Ontario	1975

large-span arches, a reentrant type of arch is used; details of a few of these structures are given in Table 1.7, in which the largest span shown is 15.29 m. A soil-steel bridge with an even larger span has been successfully constructed in Ontario in recent years; details of it are given subsequently.

The Cheese Factory bridge, which has an arch conduit, was constructed in 1984 with a record span of about 18 m. The cross section of the structure, which is located in Wellington County in Ontario, is shown in Fig. 1.27. The rise of the structure is 7.36 m and the length of the pipe 45.1 m. The conduit wall is 7 mm thick and is reinforced by longitudinal thrust beams, the locations of which are shown in Fig. 1.27, and by transverse stiffeners which are formed out of W250 × 73 rolled steel sections. The transverse stiffeners, which extend from one thrust beam to another, are spaced at 1.83 m center to center. A photograph taken during the early stages of the backfilling operation is presented as Fig. 1.28.

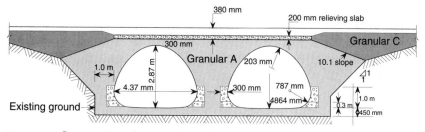

Figure 1.25 Cross section of a twin pipe-arch structure in Ontario.

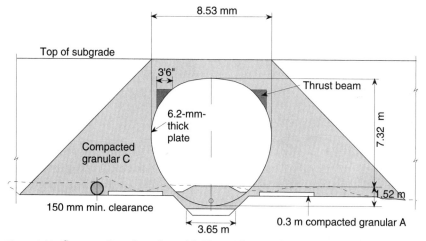

Figure 1.26 Cross section of a soil-steel bridge with pear-shaped conduit.

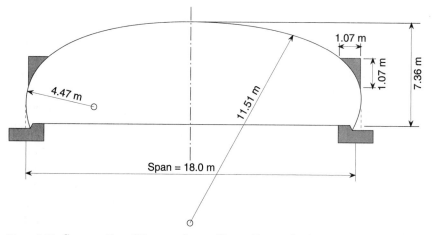

Figure 1.27 Cross section of the record span Cheese Factory bridge.

TABLE 1.7 **Examples of Soil-Steel Bridges with Reentrant Arches**

No.	Span, m	Rise, m	Depth of cover, m	Conduit wall thickness, mm	Special features	Location	Year of construction
1	15.29	5.49	1.5	6.2	longitudinal & transverse stiffeners	Norfolk County	1971
2	12.19	6.38	1.5	5.4 top segment, 4.2 the rest of the pipe	longitudinal & transverse stiffeners	Elgin County	1967
3	11.50	5.52	2.0	6.2 top segment, 4.7 the rest of the pipe	longitudinal & transverse stiffeners	Stouffville	1974
4	7.54	2.90	0.6	4.2	twin conduit	Peterborough	1975
5	7.25	3.68	1.2	4.2		Hagar	1977

The Cheese Factory bridge was load-tested after the completion of construction and was found to have more than ample strength for carrying highway live loads.

A longitudinal section through the conduit of the Cheese Factory bridge is shown in Fig. 1.29. A total of about 78 tons of corrugated steel plates and about 20 tons of rolled beam stiffeners were used for this

Figure 1.28 The Cheese Factory bridge under construction.

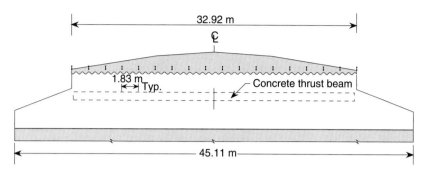

Figure 1.29 Longitudinal section through the conduit of the Cheese Factory bridge.

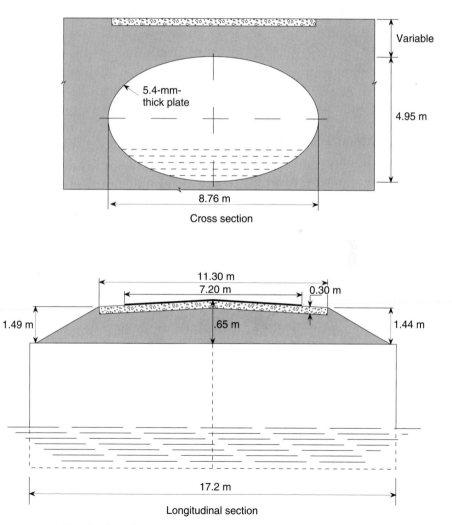

Cross section

Figure 1.30 Details of a soil-steel structure with a relieving slab.

structure. It is believed that this structure was about 20 percent cheaper than the conventional slab-on-girder type of bridge would have been.

1.4.6 A structure with a relieving slab

The longitudinal and transverse sections of a soil-steel bridge with a horizontally elliptical conduit and a reinforced concrete relieving slab are shown in Fig. 1.30. It can be seen that the depth of cover of this structure is relatively shallow. This structure was tested under vehicle loads. As will be described in Chap. 7, the relieving slab, by dispersing the concentrated wheel loads of the test vehicle, helps to reduce the live-load thrust in the conduit wall by about 50 percent.

Chapter 2

Durability of the Metallic Shell

Baidar Bakht

Ministry of Transportation of Ontario
Downsview, Ontario, Canada

Leslie G. Jaeger

Technical University of Nova Scotia
Halifax, Nova Scotia, Canada

The literal meaning of the word *durability* is capability of withstanding decay and wear. In the context of soil-steel bridges this word implies only the capability of the metallic shell of the structure to resist the loss of section due to either, or both, corrosion or abrasion. Since the incidence of abrasion in soil-steel bridges is not very common, this chapter deals only with corrosion.

The objective of the chapter is to familarize the "nonexpert" engineer with the fundamentals of corrosion of buried steel plate structures. The material presented is compiled from the wealth of information contained in contemporary relevant technical literature. Although relevant references are not cited in the main body of the text, they are listed at the end of the chapter.

2.1 Fundamentals of Corrosion

Only copper and the noble metals (namely, gold, platinum, and silver) exist in nature in their metallic forms. All other metals are found in nature in the form of their respective oxides or salts from which they are extracted, the extraction process involving consumption of energy in the form of heat. Unless protected adequately from the environment, these latter metals revert, through the process of corrosion, from their unstable metallic state to the stable natural state. When metals revert to their oxide state, the energy absorbed during their extraction is released in the form of electrical energy. Consequently, metallic corrosion is nearly always an electrochemical process. This is discussed in the following with particular reference to iron.

2.1.1 Chemistry of corrosion

It is well known that when a strip of iron is dipped into a dilute acid, such as hydrochloric acid, the metal "corrodes," leading to the formation of the soluble ferrous chloride $FeCl_2$ and hydrogen gas H_2, so that the equation of chemical reaction can be written as

$$Fe + 2HCl_2 \rightarrow FeCl_2 + H_2 \uparrow \qquad (2.1)$$

However, metal can also corrode in neutral and alkaline solutions provided that these solutions contain dissolved oxygen. Rusting is the common and familiar form of corrosion for iron; it occurs when oxygen from the air is dissolved in the moisture or water coming into contact with the metal. The equation for this chemical reaction, which leads to the formation of ferric hydroxide, is

$$4Fe + 6H_2O + 3O_2 \rightarrow 4Fe(OH)_3 \qquad (2.2)$$

Upon drying, ferric hydroxide decomposes into water and red-brown iron oxide rust, Fe_2O_3, as follows

$$2Fe(OH)_3 \rightarrow Fe_2O_3 + 3H_2O \qquad (2.3)$$

It is interesting to note that hematite, which is the most common naturally occurring iron ore, has the same chemical composition as rust.

2.1.2 Electrochemistry of corrosion

An electrochemical reaction is one in which there is a transfer of electrically charged atoms known as *ions*. When the ions are taken away from an element, the reaction is known as the *anodic reaction;* it is also called *oxidation,* although oxygen is not always involved in the process. The reaction in which ions are added is known as the *cathodic reaction,* or *reduction.*

The process of electrochemical reaction can be explained by examining the corrosion reaction in detail with reference to the reaction defined by Eq. (2.1). Noting that hydrochloric acid and ferrous chloride are ionized in water solution, this equation can be rewritten as

$$Fe + 2H^+ + 2Cl^- \rightarrow Fe^{+2} + 2Cl^- + H_2 \uparrow \qquad (2.4)$$

It is obvious that chloride which appears on both sides of the equation is not altered by corrosion. Consequently, Eq. (2.4) can be simplified by deleting the nonreacting chloride ion.

$$Fe + 2H^+ \rightarrow Fe^{+2} + H_2 \tag{2.5}$$

Equation (2.5) can be divided as follows into anodic and cathodic reactions:

$$Fe \rightarrow Fe^{+2} + 2e, \text{ anodic reaction} \tag{2.6}$$

$$2H^+ + 2e \rightarrow H_2 \uparrow, \text{ cathodic reaction} \tag{2.7}$$

The corrosion reaction, which consists of at least one oxidation and reduction reaction such as the one defined by Eqs. (2.6) and (2.7), can take place only if all the following basic elements are present. Please note, however, that the presence of these elements, while being necessary, is not itself sufficient to start the process of corrosion.

1. An anode, i.e., a metal surface where corrosion occurs
2. A cathode, which is usually another metal surface
3. A medium called the *electrolyte,* which may be water or moist soil through which ions can travel
4. A medium which is capable of conducting electricity from the anode to the cathode

These four elements are present in a typical galvanic cell which may be formed by immersing a strip of copper and one of steel in water and connecting the strips by electrical wire. As illustrated in Fig. 2.1, in this galvanic cell, the steel strip, the copper strip, the water, and the connecting wires are respectively the anode, cathode, electrolyte, and connecting medium. In the corrosion process, the current flows from

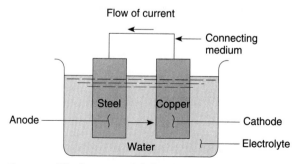

Figure 2.1 The four essential elements of a galvanic cell.

the anode through the electrolyte to the cathode because of the potential difference between the two metallic surfaces. The direction of the flow of current is always from a surface of higher potential to one of lower potential. Table 2.1 lists the various metals in descending order of their electrode potential. It can be seen in this table that copper has a lower electrode potential than steel or iron. Accordingly, for the arrangement shown in Fig. 2.1, the steel strip becomes the anode and thus corrodes. If the copper strip were replaced by a zinc strip, whose potential is higher than that of steel, the steel strip would become the cathode and the zinc strip would corrode.

The electrode potential is a measure of the energy required to extract a metal from its ore. The higher the energy required to extract a metal from its ore, the higher is the tendency of the metal to revert to its native state. The position of a metal in the kind of order presented in Table 2.1 reflects its tendency to corrode. The higher the position of a metal in this table, the higher is its tendency to corrode. For example, it can be seen that zinc is higher in this table than iron and is, therefore, more prone to corrosion that iron.

The anode and cathode do not have to be of different metals for the corrosion process to take place. Different portions of the surface of the same metallic piece can act as anodes and cathodes because of a difference in their respective potentials, which may be caused by a number of factors, such as local differences in the composition of the metal or the electrolyte and differences in residual stresses and concentration of oxygen. In this case, as shown in Fig. 2.2, the metallic piece itself also becomes the connecting medium. Of course, for the corrosion to take place, there must be an electrolyte, which can be the soil (as shown in this figure).

The difference in the electrode potential between two electrodes is responsible for the corrosion reaction. If such a difference did not exist,

TABLE 2.1 Some Metals Arranged in Descending Order of Their Electrode Potential

Serial no.	Metal	Order for electrochemical reaction	Energy consumption for extraction from ore	Tendency to corrode
1	Aluminum	↑ Anodic	↑ Most	↑ High
2	Zinc			
3	Iron			
4	Nickel			
5	Tin			
6	Lead			
7	Copper			
8	Silver			
9	Platinum			
10	Gold	↓ Cathodic	↓ Least	↓ Low

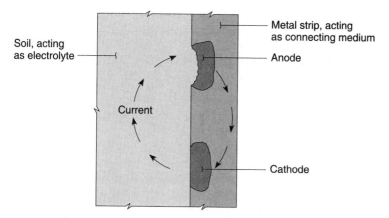

Soil, acting as electrolyte

Metal strip, acting as connecting medium

Anode

Current

Cathode

Figure 2.2 Corrosion of metal in contact with soil.

corrosion would not take place. It should be noted, however, that in practice it is practically impossible to create a condition in which the potential difference can be eliminated even from the surface of a component of the most purified metal. This is so because even this highly purified metal is nonhomogeneous at the microlevel and is likely to have different residual stresses in different regions. These conditions are responsible for the potential difference.

2.1.3 Relative tendency of metals to corrode

It has already been noted from Table 2.1 that some metals are more prone to corrosion than others. For example, it can be concluded from this table that for a given set of common factors, aluminum is likely to corrode faster than steel. Such a conclusion, while being correct for initial corrosion, may be misleading as regards the long-term performance of the metals.

The initial corrosion products form an invisible and highly protective film on aluminum, as a result of which this metal proves to be more durable than would be presumed from its position in Table 2.1. The protective film on aluminum surfaces, which is usually thinner than 0.1 mm, is the main reason why buried aluminum plate products are found to be more durable than their galvanized steel counterparts.

The protective coating formed by chemical or electrochemical reaction on the metallic surface can indeed be extremely thin. For example, in stainless steel this film, to which the steel owes its corrosion resistance, can have a thickness of only a few molecules. The rust layer on ordinary steel is also protective in nature; however, it does not stop the ingress of oxygen to the metallic surface as effectively as the protective coating on stainless steel.

2.2 Types of Corrosion Cells

There are several kinds of known corrosion cells, including galvanic cells which themselves can be divided into several categories: impressed current cells, stress cells, surface film cells, etc. Some of the more important corrosion cells associated with buried metal structures are discussed subsequently.

2.2.1 Dissimilar metal cells

A galvanic corrosion cell may consist of two dissimilar metals, as can be seen in the example of Fig. 2.1. Clearly, the greater the difference in the potentials of the two metals, the faster will be the corrosion. For example, with reference to Table 2.1, it can be readily concluded that a combination of copper and zinc will corrode much more slowly than the combination of copper and aluminum.

The incidence of corrosion because of the presence of dissimilar metals is not common in soil-steel bridges. However, it can take place if, for example, an existing pipe is extended by joining to it a length of pipe of a different metal.

The process of galvanization of steel plates which involves the joining of two dissimilar metals (discussed later in some detail) is not of concern from the standpoint of corrosion. This is because the two metal components do not have between them the electrolyte which is essential for the corrosion reaction to take place.

2.2.2 Concentration cells

One of the factors affecting the electrode potential between two regions on a surface is the composition of the electrolyte. When the composition of the electrolyte varies within the body of a component of even the same metal, a "concentration" cell may be formed. For example, the soil above the pipe may have higher concentrations of deicing salts in regions near the ends of the pipe than near its center. In such cases, as shown in Fig. 2.3, the portions of the pipe lying under the soil with the

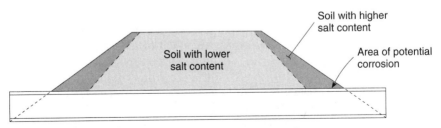

Figure 2.3 An example of corrosion through concentration cells.

higher concentrations of salt may undergo corrosion. However, it should be noted that the portion of the metal that is close to the soil with the higher concentration of salt acts as the anode only if the salt is of a metal other than the metal being corroded. If it is not, the situation reverses, causing the portion of the metal in contact with the soil having lower concentrations of the salt to act as anode.

2.2.3 Oxygen concentration cells

The most common form of corrosion in soil-steel bridges takes place due to the presence of oxygen concentration cells. Concentration of oxygen in the electrolyte at some locations and not at others sets up the oxygen corrosion cells. The portions of the metal that are in contact with the electrolyte without oxygen concentration act as anodes and corrode.

As shown in Fig. 2.4, the invert of the pipe of a soil-steel bridge rests close to the original ground, which is nearly free of oxygen. By contrast, the remaining portion of the pipe is in contact with the engineered backfill, through which oxygen can diffuse partly because of its porosity and partly because of the shortest path that the air can take. The consequence of this situation is that the invert of the pipe acts as an anode and corrodes. This explains why the inverts of soil-steel bridges are more prone to corrosion than the other portions of the pipe.

Under shallow covers the pavement can sometimes impede the ingress of oxygen, with the consequence that the pipe directly below the pavement is in contact with nearly oxygen-free soil, in contrast with the remainder of the top portion. In this case, the pipe directly under the pavement becomes the anode and corrodes.

In the top layers of the water in the pipe, oxygen concentration is caused by wave action and is responsible for the corrosion of the pipe near the waterline.

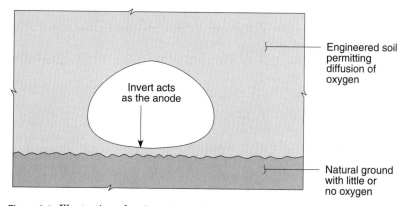

Figure 2.4 Illustration of a situation conducive to corrosion by oxygen concentration cells.

2.2.4   Temperature cells

Differences in temperature in the various portions of the pipe can also set up corrosion cells. A warmer portion of a metallic component, because of having higher stored energy, has a higher electrode potential than a colder portion. The difference in potential between the warmer and colder portions can be responsible for corrosion in which the warmer portion acts as the anode and corrodes. However, this kind of corrosion, which may readily occur in pipelines that are buried for only part of their length, is not of consequence in soil-steel bridges.

2.2.5   Impressed current cells

The types of corrosion discussed so far take place by the generation of electric current within the cell itself. Another type of corrosion can occur in which the electric current is supplied from an outside source such as a distant generator, a direct current transmission line, an electric railway, etc. Accidentally discharged impressed current can flow through the soil and find the metallic pipe in its path. In this case, the current is received in one area of the pipe and discharged from another. The latter area acts as the anode where corrosion takes place.

The impressed current can also be provided deliberately as a means of protecting the structure, as is shown in the case of cathodic protection of a structure threatened by corrosion.

2.2.6   Stress cells

The extra energy stored in a stressed metallic component raises its electrode potential. As a consequence of this phenomenon, a potential difference exists between the areas of a component that are differently stressed. The resulting potential difference can lead to corrosion, provided, of course, that all of the other conditions necessary for the setting up of a corrosion cell are present. It can be appreciated that, since the area with the higher stress will have a higher potential, the current will flow from this area to the area of lower stress, thus causing the higher-stressed area to corrode.

The corrugations for the corrugated steel plates used in soil-steel bridges are cold-formed and, as a result, residual stresses are locked into the plates. As shown in Fig. 2.5, some portions of the corrugation profile are straight, while others are curved. This situation leads to locked-in stresses being higher in the curved portions than in the straight portions, thereby creating a difference in potential which initiates the corrosion process. It can be appreciated that the curved portions of the corrugation profile, by virtue of having a higher electrode potential, act as anodes and thus corrode.

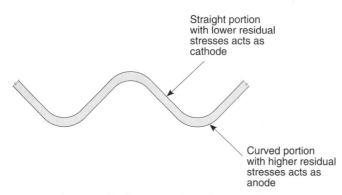

Figure 2.5 An example of corrosion through stress cells.

2.3   Factors Affecting Corrosion

From the discussion presented in Sec. 2.2, it is obvious that, except for insulating the pipe in some manner from the electrolyte, it is not a practical proposition to eliminate all those factors which are responsible for the corrosion of the metallic shell of a soil-steel bridge. All unprotected metallic pipes must corrode, although their rate of corrosion can be slowed to an acceptable level by the incorporation of appropriate factors.

Figure 2.6 is a photograph of a galvanized corrugated steel plate pipe which was removed from the ground after more than 50 years of service. This pipe bears no visible sign of corrosion. At the other extreme, there are cases in which a galvanized steel pipe has disintegrated almost completely after only about 20 years of service. Such a contrast in the durability of the pipe indicates the presence of factors which may affect the rate of corrosion of buried pipes. Some of these factors are:

- Aggressive ions
- pH of water and soil
- Composition of the backfill
- Composition of steel
- Galvanization

These factors are discussed in some detail following.

2.3.1   Aggressive ions

The most important single factor responsible for accelerating metal corrosion is the presence of aggressive ions in the electrolyte, i.e., the

Figure 2.6 A buried corrugated steel pipe after about 50 years of service.

backfill and water in the immediate vicinity of the pipe. The chloride ion is the commonly occurring aggressive ion, being present in deicing salt, acid rain, and sea water. Industrial effluent discharge can also contain aggressive ions. There is apparently no practical way of neutralizing the aggressiveness of these ions.

2.3.2   pH of water and soils

The pH value is defined as the logarithm of the reciprocal of hydrogen ion concentration in a solution. A solution with a pH of 7.0 is neutral; pH values less than 7.0 indicate an acidic solution, and those greater than 7.0 an alkaline solution. The pH value of the water and soil is still the most commonly used indicator of potential for corrosion of a galvanized steel pipe, even though several extensive observations have proven it to be inconclusive on its own.

In the distinctly acid range, for which the pH is less than about 4.0, the corrosion takes place through the kind of reaction which is defined by Eq. (2.1). For such reactions, which can take place without the presence of oxygen, the acidity of the water and backfill is the most dominant factor for corrosion. However, when the pH value is between 5.5 and 8.5, as it is for natural water, factors other than the pH value become dominant. In these circumstances, it is not prudent to rely solely upon the pH value of the water and soil as an indicator of corrosion potential.

Notwithstanding the uncertainty about the degree of influence of pH value on pipe corrosion, there is agreement between most experts that a continuously flowing stream of low-pH water constitutes a condition which is highly conducive for corrosion. It should, however, be noted that a soil with the same low value of pH provides a less severe exposure for corrosion. This is so because, unlike water, the replenishment into the soil of those agents which cause it to have a low value of pH will normally take place only gradually.

2.3.3 Composition of the backfill

It is well established that the amount of salts present in the backfill has a profound influence on the rate of corrosion of the buried metallic pipe. The salts which can enhance the corrosion reaction are those which can ionize in solution readily. An easy way to determine the approximate amount of these salts present in a soil sample is by measuring its electrical resistivity. Higher amounts of salts lead to lower resistivities. Thus, it can be concluded that a soil with low electrical resistivity is more corrosive to the metallic pipe than a soil with high resistivity.

Table 2.2 lists the typical ranges of electrical resistivity of various kinds of soils. It also indicates the tendency of the various soils to induce corrosion in a buried metal pipe. From this table it can be seen that clays and loams, because of having higher contents of salts which ionize readily in solution, provide a more corrosive environment for the metallic pipe than do sands and gravels. It is fortunate that the soils of the former group, for structural reasons that are discussed at several places in the book, are not preferred for the backfill of soil-steel bridges. Durability of the pipe provides yet another reason for not using these soils in the engineered backfill.

2.3.4 Composition of steel

There have been claims that the minor variations in the composition of steels used for corrugated metal plates, introduced over a number of

TABLE 2.2   Classification of Soils with Respect to Their Tendency to Induce Pipe Corrosion

Type of soil	Range of electrical resistivity, ohm-cm	Salt content	Tendency to induce pipe corrosion
Clay	750–2,000	↑ High	↑ High
Loam	3,000–10,000		
Gravel	10,000–30,000		
Sand	30,000–50,000		
Rock	50,000–∞	↓ Low	↓ Low

years, have enhanced the durability of the steel plates. Corrosion specialists, however, have concluded that these variations, while being useful with respect to the strength and workability of the steels, have not enhanced appreciably their resistance to corrosion.

Nearly all steels corrode rapidly in acidic water having a pH of less than about 4.0, because the corrosion product does not form a protective coating. In the pH range of 4.0 to 9.5, the corrosion product is ferric hydroxide; this forms a protective coating which retards the diffusion of oxygen, and hence slows the rate of corrosion.

2.3.5 Galvanization

Galvanizing consists of the application of a thin layer of zinc to the steel component by hot dipping. Between the outer layer of zinc and the steel base, a thin layer of a zinc-iron alloy is formed which is also protective in nature.

The zinc layer is regarded as a sacrificial coating because it protects the steel by itself corroding. It can be appreciated that the protection offered by the zinc coating is a direct function of its thickness. Most specifications require a 610 g/m^2 (2 oz/ft^2) zinc coating for galvanization of corrugated steel plates.

Galvanization was introduced in 1907 for protecting buried corrugated steel pipes. Ever since, it has been consistently used in this application and has proved to be an effective means of protection. It should be noted, however, that in acidic, or highly alkaline environments which have pH greater than 12.0, galvanization is not an effective means of protection because zinc then reacts rapidly to form soluble zinc compounds.

2.4 Recommendations

The most directly useful source of information regarding the durability of a proposed soil-steel bridge at a given location is the condition of similar existing structures and even small diameter corrugated steel pipe drains in the same general area. Of course, it is essential that the existing structures should be exposed to environments similar to those which the proposed structure is expected to experience. If the existing structures have not corroded extensively, it can be justifiably assumed that the proposed structure will behave similarly. On the other hand, if existing structures show signs of extensive corrosion, a soil-steel bridge should be avoided altogether.

In practice, only two effective measures have been used to increase the durability of the galvanized steel plate pipe of a soil-steel bridge. One of these is to ensure that the engineered backfill is of granular soil that is free of salts and organic matter; the other is to increase the thickness of the plate so that the loss of its section at the end of the anticipated life of the structure is within acceptable limits. Another effective method of protecting the metallic shell against corrosion is through sacrificial cathodic protection. In this protective system the steel plate is connected through an electrical conductor to a sacrificial zinc, or aluminum, plate which acts as anode and corrodes, thereby preventing the steel plate from corrosion.

It is essential that those situations which can lead to corrosion by dissimilar metal cells and impressed current cells be avoided.

The science of corrosion is a specialized field which involves a consideration of many more aspects than those presented in this chapter. It is expected that after reading this chapter the engineer will understand the mechanics of corrosion and the factors which influence it in a general way. It is, however, still desirable to seek the advice of a corrosion expert regarding the durability and service life design of soil-steel bridges.

References

The information compiled in this chapter has been drawn from the following references.

1. American Iron and Steel Institute, *Handbook of Steel Drainage and Highway Construction,* first Canadian edition, Washington, D.C., 1984.
2. Beaton, J. L., and Stratfull, R. F., "Corrosion of Corrugated Metal Culverts in California." *Highway Research Bulletin 223,* Washington, D.C., 1957, pp. 1–13.
3. Corrugated Steel Pipe Institute, *CSP Sewer Manual,* Mississauga, Ontario, Canada, 1977.
4. Green, N. D. Jr., "Corrosion Related Chemistry and Electrochemistry." *NACE Basic Corrosion Course,* National Association of Corrosion Engineers, Houston, Texas, 1980.
5. LaQue, F. L., "Introduction to Corrosion." *NACE Basic Corrosion Course,* National Association of Corrosion Engineers, Houston, Texas, 1980.
6. *National Cooperative Highway Research Program (NCHRP) Synthesis of Highway Practice No. 50,* "Durability of Drainage Pipe," Transportation Research Board, Washington, D.C., 1978.
7. Noyce, R. W., and Ritchie, J. M., "Michigan Metal Culvert Corrosion Study," *Transportation Research Record 713,* Transportation Research Board, Washington, D.C., 1979, pp. 1–6.
8. Parker, M. E., "Corrosion by Soils," *NACE Basic Corrosion Course,* National Association of Corrosion Engineers, Houston, Texas, 1980.
9. Temple, J. W., and Cumbaa, S. L., "Evaluation of Metal Drainage Pipe Durability After Ten Years," *Transportation Research Record 1087,* Transportation Research Board, Washington, D.C., 1986, pp. 7–14.

Mechanics of Behavior

Baidar Bakht

Ministry of Transportation of Ontario
Downsview, Ontario, Canada

Aftab A. Mufti

Technical University of Nova Scotia
Halifax, Nova Scotia, Canada

Leslie G. Jaeger

Technical University of Nova Scotia
Halifax, Nova Scotia, Canada

3.1 Introduction

Despite their being in existence for several decades, soil-steel bridges are usually designed by reliance upon empirical knowledge, supported by rudimentary analyses which are based upon oversimplifying assumptions. Different experts on the subject have opinions about how these structures should be designed, and these opinions differ so widely from one to another that a "nonexpert" engineer, because of not understanding the mechanics of behavior of the structure, has no choice but to follow the advice of one expert as an act of faith.

Another consequence of the lack of understanding of the mechanics of behavior of soil-steel bridges is that when these structures are in distress the engineer, being unable to determine the cause of distress, cannot make a rational decision regarding the corrective action that must be taken. It is noted that distress in soil-steel bridges is more prevalent than in their conventional counterparts.

In spite of being inexpensive and relatively easy to construct, soil-steel bridges are quite difficult to analyze through mathematical models; this is the main reason why their mechanics of behavior is not well understood. This chapter explores the mechanics of their behavior through fairly simple mechanical models. Results of finite element analyses and field measurements are used to complement the conclusions arrived at by the study of the mechanical models.

3.2   Load Sustenance

Examples of a soil-steel bridge and its bare metallic shell during fabrication are shown in Figs. 3.1 and 3.2, respectively. The shell on its own is so flexible in bending that in order to maintain its cross-sectional shape it sometimes has to be braced by ties and struts. In fact, this shell is so weak in flexure that if it were scaled down by principles of structural modeling to the size of a soup can, it would be too flimsy to handle manually.

The deformation of a soup can with its lids removed is shown schematically in Fig. 3.3a under the action of vertical loads. It can be appreciated readily that it does not require a large load to deform the soup can as shown in this figure. The situation becomes very different if, as shown in Fig. 3.3b, the same soup can is held between two wooden blocks, each of which is shaped to the curvature of the soup can. Even when the gap between the two wooden blocks is as much as about 10 mm, this structural system can easily withstand the weight of an average man. It is not difficult to visualize that the portion of the soup can between the two blocks is predominantly in compression. Indeed,

Figure 3.1 A soil-steel bridge during construction.

Figure 3.2 The bare metallic shell of a soil-steel bridge during construction.

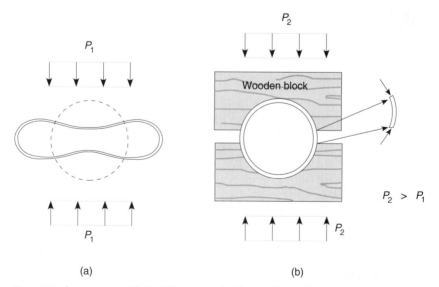

(a)   (b)

Figure 3.3 A soup can with its lids removed: (*a*) metallic shell on its own; (*b*) metallic shell between two wooden blocks.

the entire wall of the soup can along its circumference is predominantly in compression.

The soup can of Fig. 3.3*b* is able to sustain very high loads because the applied vertical loading is transformed into radial loads which compel the can to be mainly in compression. A soil-steel bridge is similar to the combination of the soup can and the two shaped wooden blocks. The gravity load of soil in a soil-steel bridge compels the metallic shell to deform laterally and press against the backfill, thereby generating the lateral pressures. The backfill soil around the metallic shell performs two distinct roles. In one role it is responsible for the load that the shell is called upon to sustain; in the other it provides the necessary support to the shell to enable it to sustain its weight.

3.2.1   Infinitely long tube in a half-space

For the analysis of soil-steel bridges it is usual to assume that the metallic shell or tube is infinitely long, and that it is buried in a semi-infinite space, or half-space, so that one transverse slice of the structure, which is shown conceptually in Fig. 3.4, is similar in behavior to any other transverse slice. The advantage of this assumption is that the very complex three-dimensional nature of the actual structure can be investigated using a relatively simple two-dimensional plane-strain idealization. It is recalled that in a plane-strain idealization, deformations perpendicular to the plane of the idealized structure are assumed to be zero.

The simplification afforded by the two-dimensional idealization is not without disadvantages, although it provides a useful tool to study the behavior of soil-steel bridges. We shall first study the behavior of soil-steel bridges with the help of this idealization, and discuss its disadvantages later.

Bending effects.   It can be shown that if a segment of a shell of varying radius of curvature *r* is subjected only to radial pressures which vary

Figure 3.4 A transverse slice of a soil-steel bridge.

inversely to *r,* as shown in Fig. 3.5, then the shell remains free from any moments, i.e., it is subjected to only thrust and shear.

In an actual structure, the radial pressures on the metallic shell do vary approximately, although not precisely, in inverse proportion to the radius of curvature. Because of the radial pressures not being exactly proportional to $\frac{1}{r}$, and because of load effects imposed during the construction process, the shell of an actual soil-steel bridge is subjected to some bending moments; these are, however, limited by the plastic moment capacity of the shell. As discussed earlier, the corrugated plates usually employed in soil-steel bridges are quite weak in flexure, because of which the plates can develop plastic hinges under fairly low moments. Such hinges are usually formed at the crown and shoulders, as shown in Fig. 3.6.

For loads applied subsequent to the formation of the plastic hinges, the corrugated metal plate shell behaves rather like a ring with a few hinges. Such a ring, even when subjected to radial pressures that are not proportional to the inverse of the radius of curvature, does not permit the formation of substantial bending moments.

In soil-steel bridges it is usual to ignore bending moments and design the metallic shell for thrust only. It should be noted, however, that the moments can be ignored only if the conduit walls are flexurally quite weak. If they are not weak, as in the case of concrete pipes or corrugated metal structures with plates of deep corrugations, and have shallow depth of cover, the moments should be taken into account in the design process.

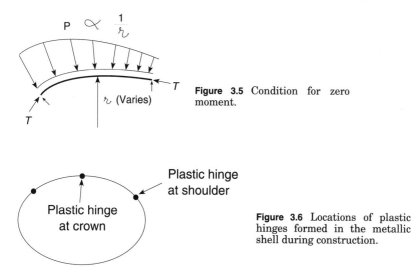

Figure 3.5 Condition for zero moment.

Figure 3.6 Locations of plastic hinges formed in the metallic shell during construction.

Although the moments in the metallic shell may not be taken into account in the design process, their effects on the integrity of the structure cannot always be ignored, as is subsequently discussed.

Excessive bending deformations in the metallic shell result mainly from the inability of the backfill to sustain the radial pressures which develop around the conduit due to the interaction of the soil and the shell. It is true that the development of substantial bending moments is inhibited by the relatively small flexural rigidity of the metallic shell and by the formation of plastic hinges at strategic locations. However, the absence of substantial bending moments does not eliminate large bending deformations of the plate; these can be very detrimental to the integrity of the structure, especially if they take place at the locations of the bolted longitudinal seams of the plate.

Pipe-arch-type structures have plates of very large radius of curvature at the bottom and of rather small radius of curvature at the lower sides, or haunches. Reference 1 shows that in these structures the radial pressure in the backfill soil is the highest under the haunches, where it is also quite difficult to compact the backfill. The net result of this situation is that the plate undergoes substantial bending in the haunch areas. This bending may either lead to crimping of the plate or to the development of large cracks in the bolt holes. Prevention of bolt hole cracking, which is caused by excessive bending of plates, is possible only by the provision of very stiff backfill adjacent to those portions of the plate where the radial soil pressures are expected to be high.

Arching effects.   If, in a soil-steel bridge, the points along every horizontal line have the same deflections due to the gravity load of the soil, then the structure is in a plane-strain condition and is composed of plane-strain slices. In such a structure no transference of load takes place from the column of soil immediately above the conduit to the adjacent columns of soil. Clearly, such a condition can occur only if the load-deformation characteristics of the embedded pipe with respect to vertical loads are exactly the same as the corresponding load-deformation characteristics of the body of soil displaced by the conduit.

The condition in which loads from one column of soil are not transferred to another is shown schematically in Fig. 3.7a; it can be regarded as the no-arching condition.

If the column of soil containing the conduit were to deflect more than the adjacent columns, as shown in Fig. 3.7b, then it would tend to drag down the two adjacent columns of soil along with it and, in so doing, would pass on some of its own load to these columns. The effect of this load transference is to relieve the pipe of some of the load that it would have sustained in the absence of the load transference. This condition in a soil-steel bridge is referred to as the *positive arching condition*.

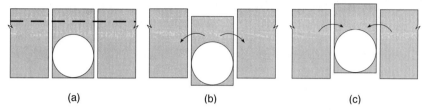

Figure 3.7 Illustration of arching effects: (*a*) no arching condition; (*b*) positive arching condition; (*c*) negative arching condition.

The condition of negative arching is clearly that in which the pipe is called upon to sustain more load than it would have sustained if there were no load transference from the column of soil immediately above the conduit to the adjacent ones. This condition, as shown in Fig. 3.7*c*, can occur when the adjacent columns of soil deflect more than the middle column, thus causing some of their loads to be transferred onto the pipe.

It is interesting to note that, strictly speaking, the no-arching condition is a fictitious one because it is virtually impossible for an embedded pipe to have exactly the same stiffness as the body of displaced soil; the latter stiffness varies from a maximum at the crown of the conduit to almost zero at the spring lines, or the outer ends of the conduit.

The positive and negative arching conditions should also be regarded only as qualitative measures; their quantification is made impractical by the fact that, as is shown later, different load effects are influenced differently by arching. For example, the effect of arching on the vertical soil pressure over the crown may be quite different from that on the vertical soil pressure near the spring lines; moreover, the thrust, which does not remain constant along the circumference, may be affected differently from the vertical soil pressures at both crown and spring line.

Finite element analyses.   To gain an insight into their behavior, soil-steel bridges with conduits of different shapes were analyzed by the finite element (FE) method by idealizing them in plane-strain. The analyses were conducted by assuming:

1. The gravity loads are "switched on" after the pipe has been placed in position in the soil.

2. The backfill soil and the foundation are of the same homogeneous linear-elastic medium, with the same stress-strain relationship in both tension and compression.

3. There is complete bond between the soil and the metallic shell at their interface.

The foregoing assumptions, which can be challenged with good reason, are discussed later. These simplistic assumptions are justifiable for the study at hand because the results of the analysis are free from the significant additional complications which would arise if they were not made. Making use of the symmetry of both the structure and the loading, only one-half of the structure was analyzed in each case. The presence of symmetry was reflected in the analysis by imposition of the condition of no horizontal movement at the vertical centerline.

Details of the four structures analyzed by the FE method are given in Fig. 3.8. It can be seen from this figure that structure I has a vertically elliptical conduit, structure II a round one, and structure III a horizontally elliptical one. The distance of the top of the embankment measured from the spring line is kept the same for structures I, II, and III. As shown in Fig. 3.8, structure IV has a horizontally elliptical conduit with the same proportions as that of the conduit of structure III, with the difference that all its dimensions are double those of structure III.

The extent of the backfill and foundation considered in the FE analysis is also shown in Fig. 3.8. It can be seen that in the horizontal direction the backfill included in the analysis extends 2.5 times the conduit span beyond the spring line, and that below the invert the foundation extends 1.5 times the rise.

The conduit wall thrusts T_D obtained by the FE analyses are nondimensionalized by dividing them by the weight W of one-half the column of soil directly above the conduit. It is noteworthy that T_D would have been equal to W if the no-arching condition existed in the structure and the thrust were uniform around the conduit circumference. The nondimensionalized values of the conduit wall thrust are plotted in Fig. 3.9 for the four structures analyzed by the FE method. To facilitate direct comparison, these values are plotted along a straight line of nondimensional length which runs from the crown of the conduit to its invert.

Similarly to the conduit wall thrusts, the vertical stresses in the soil around the conduit were also nondimensionalized by dividing them by the corresponding free-field pressures. It is recalled that the free-field pressure in the soil is the pressure which would have existed in the absence of the conduit. The nondimensionalized vertical pressures along three different horizontal planes are plotted in Fig. 3.10 along a straight line which runs along the conduit span, and which is also nondimensionalized. As shown in Fig. 3.10, this line starts at the vertical line of symmetry and extends three times the conduit span D_h. The vertical pressures are plotted in Fig. 3.10 at the crown level, at the level of the spring line, and just below the invert.

In light of the results of the FE analyses presented in Figs. 3.9 and 3.10, several interesting observations can be made regarding the

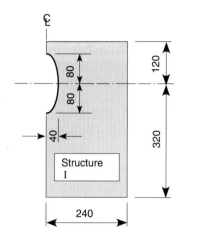

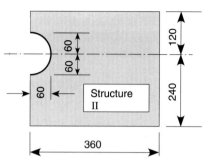

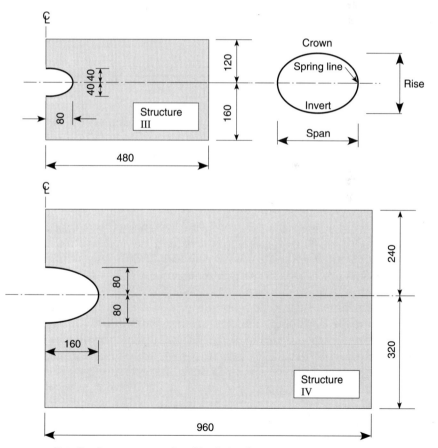

Figure 3.8 Details of structures analyzed by finite elements.

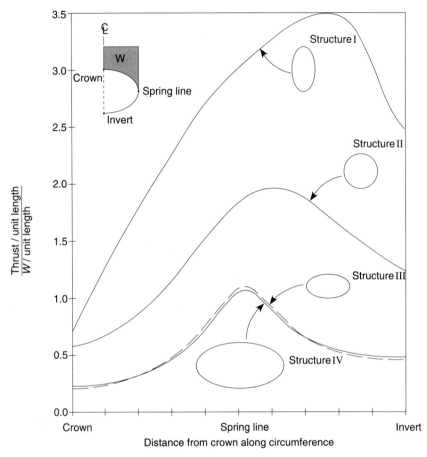

Figure 3.9 Thrust in the conduit wall around the conduit circumference.

behavior of soil-steel bridges with conduits of different shapes; these observations are presented subsequently under different headings.

Nonuniform thrust.   It can be seen in Fig. 3.9 that the thrust in the conduit wall is far from being uniform around the conduit circumference. To enable better visualization of the thrust pattern, the nondimensionalized values of thrusts in structures II and III are replotted in Fig. 3.11 around half the circumference of the respective conduits. It can be seen in these figures that in the structure with a circular conduit the thrust has a maximum value just below the spring line, and in the structure with a horizontally elliptical conduit the maximum thrust occurs at the spring line. It is noted that these observations do not conform with field measurements, an example of which is shown later. Field measurements of the response of the soil-steel bridges dur-

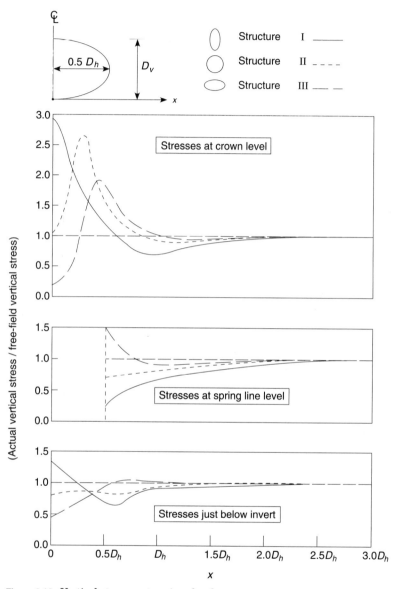

Figure 3.10 Vertical stresses at various levels.

ing and after the construction indicate that the maximum thrust
occurs about midway between the crown and the spring lines.

The lack of correlation between the predictions of the FE analyses
and field measurements may be attributed mainly to the fact that the
effects of construction sequence were not included in the analysis.
Notwithstanding this lack of correlation, the FE analysis clearly

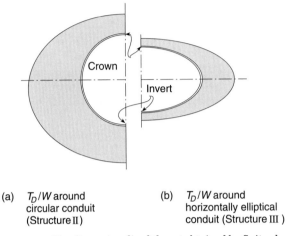

(a) T_D/W around
circular conduit
(Structure II)

(b) T_D/W around
horizontally elliptical
conduit (Structure III)

Figure 3.11 Nondimensionalized thrust obtained by finite element methods.

demonstrates the presence of significant nonuniformity of the thrust around the conduit. It can be readily appreciated that when the flexural rigidity is relatively small, the nonuniformity of thrust around the conduit can be maintained only by the presence of interface shear between the backfill and the conduit wall. When the backfill around the conduit consists of noncohesive granular soil, as it ideally should, the interface shear between the conduit wall and backfill can develop only through friction between the two.

Extreme examples of the nonuniformity of thrust around the conduit wall are found in thrusts induced by vehicle loads on the embankment above the conduit. One such example is presented in Fig. 3.12, which shows thrusts in the conduit wall computed from measured strains which were induced by a test vehicle placed on one side of the crown, as is also shown in the figure. It can be seen that the portions of the conduit wall which are remote from the vehicle load experience hardly any thrust due to the vehicle weight. The results presented in Fig. 3.12 are from a test not yet reported.

Effect of arching on thrusts. Figure 3.9 clearly shows that the nondimensionalized thrust is the highest in the structure with the vertically elliptical conduit, with the ratio T_D/W being as high as about 3.5. It is recalled that the no-arching condition corresponds to $T_D/W = 1.0$. The fact that T_D/W is significantly greater than 1.0 suggests that the embedded pipe of this structure is considerably stiffer than the soil which it replaced, resulting in significant negative arching. The presence of negative arching is also confirmed by the very high vertical soil pressure at

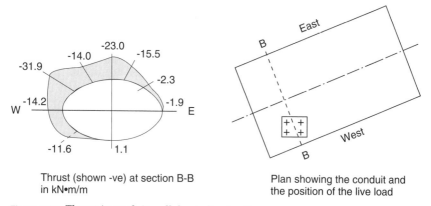

Thrust (shown -ve) at section B-B
in kN•m/m

Plan showing the conduit and
the position of the live load

Figure 3.12 Thrust in conduit wall due to live loading.

the crown in the structure under consideration. This very high pressure at the crown is shown in Fig. 3.10, in which it can be seen that the vertical soil pressure above the vertically elliptical conduit is about three times the corresponding free-field pressure. Clearly, the vertically elliptical shape of conduit attracts load to the metallic shell.

The values of T_D/W in the structure with a horizontally elliptical conduit are the smallest of the three structures with conduits of different shapes. As can be seen in Fig. 3.9, the value of T_D/W remains well below the no-arching value of 1.0, except near the spring line where it is close to 1.0. It is obvious that a horizontally elliptical conduit generates at least some positive arching of the kind shown schematically in Fig. 3.7b. Because of this positive arching condition, much of the load of the column of soil directly above the conduit is transferred to the adjacent columns of soil.

As can also be seen in Fig. 3.9, the value of T_D/W in the structure with a circular conduit varies between about 0.6 and 2.0, indicating that this conduit shape can also generate negative arching, but of a smaller magnitude than that generated by the vertically elliptical conduit. This result accords with what engineering judgment would suggest.

Vertical pressure at crown level.   The variations of vertical soil pressure at the crown level shown in Fig. 3.10 are both interesting and instructive. It can be seen that for a structure with a vertically elliptical conduit this pressure peaks immediately above the crown and then rapidly diminishes to the free-field pressure at a position which is vertically above the spring line. For a structure with a circular conduit, the vertical soil pressure is equivalent to the free-field pressure at the crown; it peaks about midway between the crown and spring line and then drops to the free-field pressure about one-third of the span length away from the spring line.

In the case of a structure with a horizontally elliptical conduit, the vertical soil pressure above the crown is about 20 percent of the free-field soil pressure. As one moves away from the crown, the pressure increases and reaches a peak at a station which is within the conduit cross section but close to the plane containing the spring line; it then decreases to the free-field pressure at about one-half span length away from the spring line. The fact that the "hard point" which attracts the peak vertical stress lies within the conduit cross section confirms that the stiffness of the embedded tube near the spring line is considerably more than the corresponding portion of the soil which is displaced by the conduit.

Vertical soil pressures immediately above the conduit have often been regarded as indicative of the presence of positive or negative arching effects. Positive arching conditions can be said to exist when these pressures are below the free-field pressures, and negative when they are above. It can be seen in Fig. 3.10 that, while in structure I the condition of negative arching exists across the entire cross section of the conduit, this is not the case for the other two structures. In these two, part of the cross section of the conduit is in the positive arching condition and part in the negative arching condition. This observation confirms that it is not feasible to quantify arching for the whole of a structure by means of a single factor.

Vertical pressure below the conduit. Nondimensionalized vertical stresses in the foundation just below the invert level are also plotted in Fig. 3.10 for structures I, II, and III. It can be seen that for structure I with the vertically elliptical conduit, the vertical stress just below the invert is about 20 percent higher than the corresponding free-field pressure. This observation may appear contrary to engineering judgment, according to which it would appear that the stresses under a cavity in a half-space should always be somewhat smaller than the corresponding stresses in the absence of the cavity.

In the structures with circular and horizontally elliptical conduits, the vertical soil pressures under the conduit are smaller than the corresponding free-field pressures. As discussed later, these smaller pressures under the conduit can create an undesirable situation if the foundation on which the structure is built is too yielding.

Scalar effects. As can be seen in Fig. 3.9, the nondimensionalized thrusts in structures III and IV are almost exactly similar, although the latter structure is twice as big as the former. This observation confirms that for the conditions assumed in the analysis, the pattern of behavior of a soil-steel bridge, while being affected by the shape of the conduit, is not influenced by the overall size of the structure. If two structures of similar shapes but different sizes are found to behave differently in practice, then the reason for such a difference should be sought in factors other than those which are already implicit in the FE analyses discussed earlier.

3.2.2 Three-dimensional effects

The results of analyses discussed so far relate only to two-dimensional idealizations in which it is assumed that one transverse slice of the structure behaves exactly like any other. In reality, this assumption is not borne out because the fill above the conduit is of uniform depth only above the middle portion of the conduit length. As shown in Fig. 3.13, the end portions of the pipe are subjected to backfill of decreasing height.

Recognizing that the foundation below the middle portion of the tube, because of being subjected to deeper fills, is likely to settle more than the foundation below its outer ends, the pipe is usually laid with an exaggerated camber in the middle, as shown in Fig. 3.14. The camber is so adjusted that, after settlement, the pipe eventually lies in the designed position; in the case of a culvert, this designed position will frequently be an inclined rather than a horizontal one.

Figure 3.13 Longitudinal section through the pipe.

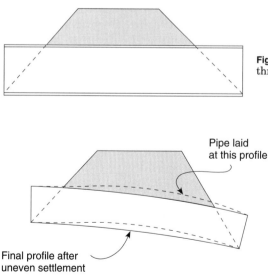

Pipe laid
at this profile

Final profile after
uneven settlement

Figure 3.14 Correction for uneven settlement along the length of the pipe.

It is usual not to change the size of the plates along the length of the conduit, despite recognition that the middle part of its length is subjected to heavier loads than the outer parts. Mainly because of this practice, it is considered safe to take the analysis of the middle portion of the pipe as being valid for the whole length. Another reason for ignoring the effect of uneven loading along the pipe length is that the pipe, with its annular corrugations, is perceived to have practically no overall flexural rigidity in the longitudinal direction. This perception is quite valid for the pipe considered on its own, in which case (as shown in Fig. 3.15*a*) the pipe can deform longitudinally like the bellows of an accordion, with minimal bending moments. However, when this flexible pipe is embedded in soil as shown in Fig. 3.15*b*, the free longitudinal movement of the corrugation rings is restrained by the soil; in this case, the pipe no longer remains as flexible as it is when considered in isolation.

It can be appreciated that when the longitudinal flexural rigidity of the embedded pipe is not negligible, the effect of uneven loading along its length cannot be ignored.

Longitudinal arching.   If the pipe deflects unevenly along its length, as shown in Fig. 3.16, then it can be foreseen that some of the load above it will be transferred from the middle portions of the pipe to the outer portions, thereby relieving the conduit wall of the middle portion of some of its load effects. This transference of load in the longitudinal direction of the conduit can be regarded as a consequence of longitudinal arching, as distinct from the arching discussed earlier in relation to the plane-strain idealization of the structure; for convenience, this latter can be referred to as *transverse arching.* It is noted that because of difficulties in analyzing the structure in three dimensions, very little work has been done to date to study the effects of longitudinal arching.

Another consequence of longitudinal arching is that when the top portions of the conduit walls are stiffened by transverse stiffeners,

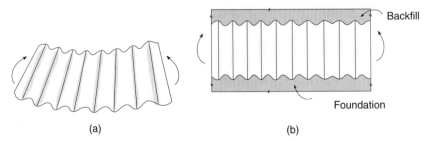

Figure 3.15 Illustration of the longitudinal flexural rigidity of the corrugated plate pipe: (*a*) corrugated plate pipe on its own; (*b*) embedded corrugated plate pipe.

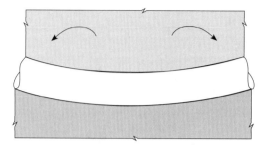

Figure 3.16 Illustration of longitudinal arching.

which may be spaced apart by up to 4 m, the soil load may be transferred to these stiffeners in the manner shown in Fig. 3.17. The stiffeners may consist of either curved rolled I-beams, or curved corrugated
plates placed in a ridge-over-ridge fashion, as can be seen in the structure shown in Fig. 3.18. Both kinds of stiffeners are bolted to the
erected shell of the structure through holes which are usually made by
welding torches. Because of the imperfect connections resulting from
this process, the transfer of shear between the stiffeners and the parent plates is not always perfect, especially during the early stages of
backfilling above the conduit.

The effect of longitudinal arching between the stiffeners is clearly to
relieve the portion of the conduit wall between the stiffeners of some
soil weight and to transfer it to the stiffeners.

Effect of foundation settlement.   Ideally, the material which should be
used for the backfill around the conduit is well-compacted granular
material; for all practical purposes, this material has no time-dependent
properties. The foundation of the structure, on the other hand, may not

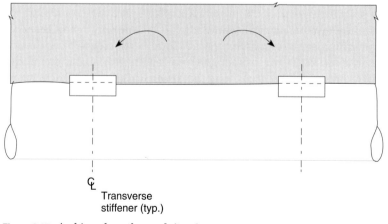

Transverse
stiffener (typ.)

Figure 3.17 Arching along the conduit axis.

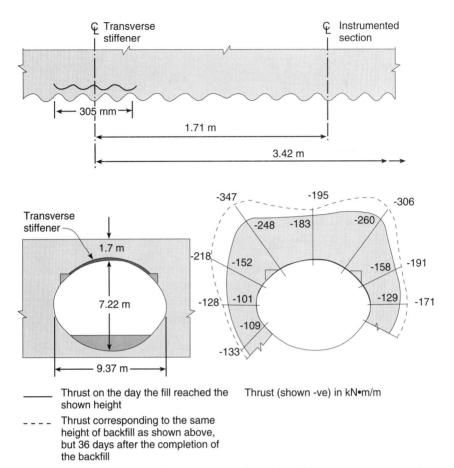

Figure 3.18 Thrust in the conduit wall at a section midway between transverse stiffeners, computed from measured strains.

be of the same quality as the backfill. If the foundation is composed of soil of a predominantly cohesive type, its total settlements may consist both of those which occur immediately after the construction of the structure and those which occur long after; the latter may in some cases continue to increase for a long time. Both kinds of settlements have their influence on longitudinal arching and transverse arching.

As discussed in the subsection immediately preceding, and shown schematically in Fig. 3.14, the foundation under the conduit can settle unevenly in the longitudinal direction of the pipe. In addition, there is likely to be uneven settlement in the transverse direction as well. The nondimensionalized vertical soil pressures below the invert level, plotted in Fig. 3.10, can conveniently be used to study uneven settlement of the foundation in the transverse direction.

In the case of structure I, with the vertically elliptical conduit, the vertical soil pressures under most of the conduit are much higher than the pressures at the same level in adjacent portions of the foundation. It is obvious that in this case the foundation under the conduit will settle more than in the adjacent portions, and will create a positive arching condition.

Unlike structure I, structures II and III (which have circular and horizontally elliptical conduits, respectively) have much smaller vertical pressures under the conduit than the corresponding pressures away from the conduit. In this case, the long-term settlement of the foundation will create the situation shown schematically in Fig. 3.7c, which generates negative arching.

It can be seen readily that the vertically elliptical conduit, while generating a negative transverse arching condition in the time-independent backfill, may experience a positive transverse arching condition due to time-dependent settlement of the foundation. The opposite is true for the horizontally elliptical conduit, which may generate positive arching in the backfill considered in isolation, but may experience some negative arching if the structure rests upon a foundation with significant time-dependent deformation characteristics.

An example.   An example of the complex interaction of arching in the longitudinal and transverse directions is now presented. The example is that of conduit wall thrusts, computed from strains during construction, in the metallic shell of a soil-steel bridge with a horizontally elliptical conduit having span and rise of 9.37 and 7.22 m, respectively. The metallic shell of this structure is installed with a concrete longitudinal stiffener at each of the conduit shoulders, and with transverse stiffeners which extend transversely from one longitudinal stiffener to the other. Details of the structure are given in Fig. 3.18, which shows that the 305-mm-wide stiffeners are made of corrugated plates and that they are spaced at 3.42 m center to center. The depth of soil cover over the instrumented section is 1.7 m.

The structure under consideration, called the Upper Access Road bridge, was monitored during its construction by the first-named author of this chapter. The conduit wall of this structure, a photograph of which is shown in Fig. 3.2, was instrumented with strain gauges at several stations around a transverse section located midway between two adjacent transverse stiffeners in the middle length of the pipe. There were three strain gauges at each station, to enable both thrusts and moments in the conduit wall to be calculated. The strain gauges were installed soon after the metallic shell was assembled from curved corrugated plates. Readings from the strain gauges were recorded as the fill height grew on both sides of the pipe.

Soon after the fill reached the specified maximum height, readings from all the instruments were recorded. The thrusts computed from these measured strains are shown in Fig. 3.18 by a solid line. Unfortunately, the strain gauges in the lower portion of the conduit were damaged during construction, as a result of which the thrusts in the lower portions of the conduit wall could not be calculated.

As shown in Fig. 3.18, the dead-load thrusts soon after the completion of the structure are far from being uniform around the circumference of the conduit. The maximum thrust occurs at the shoulder of the conduit and not near the spring line as predicted by the FE analysis mentioned earlier. The lower portions of the conduit are subjected to much smaller thrusts than the upper portions. This outcome is also in conflict with the predictions of the FE analysis, according to which the lower portions are subjected to higher dead-load thrusts than the higher portions.

Thirty-six days after the fill above the conduit had reached the specified level, the strains in the conduit wall were recorded again. Surprisingly, it was found that the strains had generally increased, despite the fact that the structure was not open to traffic and no apparent change had taken place in the condition of the structure during this time. Thrusts computed from the latter strains are shown in Fig. 3.18 by dotted lines. It can be seen in this figure that during the 36 days since completion of the structure the dead-load thrusts had generally increased all around the circumference of the conduit. However, the increase in the conduit wall thrust over the crown region was much smaller than the increase over the other portions.

While there is no definitive explanation for this apparently strange phenomenon, it is postulated that the foundation of the structure underwent time-dependent settlements during the 36 days after completion of construction. For reasons given earlier, the settlements on either side of the horizontally elliptical conduit would have been larger than those under the conduit. This situation, again as discussed earlier, led to negative transverse arching, as a result of which the thrusts increased in the conduit wall. The overall uneven settlement of the pipe in the longitudinal direction also caused the soil above the crown to arch between the stiffeners. Since the stiffeners are contained between the two longitudinal stiffeners situated at the shoulders, the longitudinal arching was also limited to the region between the shoulders. It can be appreciated that in the shoulder area the longitudinal stiffeners would tend to keep the thrust in both the stiffened and unstiffened portion of the conduit wall fairly uniform. For this reason, the nonuniformity of thrust would take place in regions well away from the longitudinal stiffeners. This remote region from the longitudinal stiffeners is clearly the region above the crown where the thrusts did not increase as in the other portions of the conduit wall.

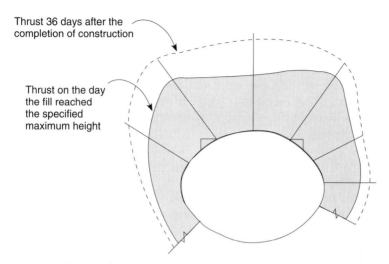

Thrust 36 days after the
completion of construction

Thrust on the day
the fill reached
the specified
maximum height

Figure 3.19 Projected thrust in the conduit wall at a section under a transverse stiffener.

Unfortunately, the section of the conduit wall under a stiffener was not instrumented. However, according to the foregoing arguments, it appears that this section would also have seen an increase in the conduit wall thrust during the 36 days after the completion of the construction, but that the pattern of increase would have been as shown in Fig. 3.19. This figure is drawn on the assumption that just after the completion of construction the dead-load thrust at a section midway between the stiffeners would have been the same as the thrust at a section containing the stiffener. In the figure it is shown that at a section containing a transverse stiffener the increase in conduit wall thrust over the crown is likely to be much more than in other portions of the conduit wall.

It is obvious that there is much that yet needs to be learned about the behavior of soil-steel bridges especially as they are affected by the time-dependent settlements of the foundations. When the foundation is such that it is expected to have time-dependent deformations, then account should be taken of its detrimental effects on the conduit wall thrust; many such structures have experienced distress despite the fact that they were constructed properly with well-compacted granular backfill and under strict and competent supervision.

3.3   Assumptions of Finite Element Analyses

The validity of the assumptions employed in the foregoing FE analyses in representing closely the behavior of the actual structure will now be discussed.

3.3.1 Plain-strain idealization

It has already been noted that the plane-strain idealization cannot be used to analyze the effects of longitudinal arching, and that these effects may be significant if the structure rests upon a relatively yielding foundation. However, when the foundation is relatively unyielding and consequently the foundation settlements are insignificant, the analysis of a soil-steel bridge by idealizing it by plane-strain slices can give realistic results.

The plane-strain idealization is also not suitable for the analysis of soil-steel bridges subjected to concentrated loads at the embankment above the conduit. This is because of the difficulty in accounting for longitudinal dispersion of the concentrated load in the idealization of the load that should be applied to the transverse slice of the structure.

3.3.2 Accounting for construction stages

The FE analyses discussed earlier were performed by assuming that the gravity load of the soil was "switched on" after the pipe was placed in position in the soil. This assumption, while appearing realistic at a cursory level, is not fully justifiable for the following reasons.

The behavior of two different pipes is contrasted. One has a cross section which will not deform appreciably under pressures imposed by the backfill, so that the side walls of the conduit remain nearly unyielding under the action of lateral earth pressure. The other structure has an easily deformable cross section so that its side walls move rather easily in response to horizontal earth pressure. The side walls of these two structures are now subjected to the first compacted layer of backfill. As shown in Fig. 3.20a, the side walls of both these structures would be subjected to similar lateral pressures which vary linearly from zero at the top of the layer to a maximum value at the bottom.

A second layer of compacted backfill is now placed above the first one. The side wall of the former structure does not yield, so that the effect of placing the second layer of the backfill is simply to increase the lateral pressure on the side wall as shown in Fig. 3.20b. It can be appreciated that for this structure the sequence of placing the backfill layers is irrelevant to the lateral pressures on the side wall.

The side wall of the latter structure yields under the second layer, causing the side wall to move away slightly from the first layer of backfill, thereby relieving some of the horizontal pressure due to the first layer. The consequence of this situation is that, as shown in Fig. 3.20b, the pressure diagram corresponding to the first layer has a steeper slope. The placement of the third layer of backfill, as shown in Fig. 3.20c, besides reducing the pressure induced by the second layer, further reduces the pressure induced by the first layer, so that the pres-

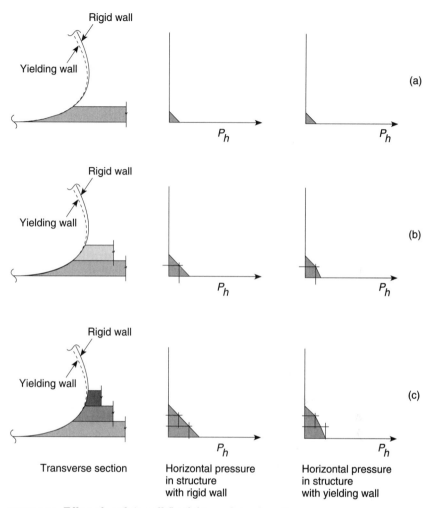

Figure 3.20 Effect of conduit wall flexibility on lateral earth pressures.

sure diagram begins to curve downward as shown in the figure. The pressure diagram corresponding to the former structure, of course, maintains the same slope for all layers of backfill.

It can be appreciated that the sequence of placing the layers of backfill on either side of the conduit has a significant influence on the load effects in the conduit wall. Clearly, this influence can be accounted for only if the sequence of construction is included in the analysis and the method of analysis is able to account for the second-order effects of the movement of the conduit wall on soil pressures.

Even when the sequence of placing the backfill layers is taken into account in the analysis, it is assumed that the layer of soil is added to the structure with the properties of the compacted backfill, so that this layer participates to an appreciable extent in sustaining its own weight. This is far from reality, as will be seen by considering the case shown in Fig. 3.21, in which the compacted fill has reached the shoulder level and the soil has been placed above the conduit to be compacted. It is obvious that all the weight of the loose soil, which does not yet have any structural strength, would have to be sustained almost entirely by the conduit wall. The compacting of this layer of soil, without changing the load effects in the conduit wall due to its own weight, will help to disperse the loads of the soil placed above it, and so on.

In an ideal situation, the load of each layer of the backfill should be distributed in the structure by assuming that it has no strength. However, when adding subsequent layers, this layer should be given the properties of compacted backfill.

3.3.3 Assumption of linear elasticity

Well-compacted granular material does indeed act in a linear elastic manner for all practical purposes. However, it may be prudent to account for the fact that the soil becomes stiffer with increase in the overburden pressure, with the consequence that the effective modulus of elasticity of the soil increases with depth. There are several models available in the technical literature which can be used advantageously in the analytical studies of soil-steel bridges.

The foundation can be treated as an elastic medium whose elastic properties are adjusted in a time-dependent way to reflect its long-term behavior. However, to assign the same properties to the foundation as to the backfill, as was done for the FE analyses discussed earlier, is realistic only if the two media have similar properties.

Contrary to the assumption made in the analyses mentioned earlier, soil does not have the same stress-strain relationship in tension as in compression. Clearly, soil is almost totally incapable of supporting

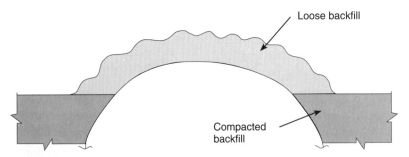

Figure 3.21 A layer of loose backfill above the conduit.

tension. The consequence of making this assumption is not very significant, however, because the backfill of a soil-steel bridge is predominantly in compression except in the soil above the conduit when the depth of cover is very shallow.

3.3.4 Bond between soil and metallic shell

In the preceding FE analyses, complete bond was assumed between the soil and the conduit wall. A more realistic idealization would have been one in which the interface shear between the conduit wall and soil could develop only through friction, with provision for relative slip between the two media if the frictional resistance was exceeded by the interface shear. It is suggested, however, that the inaccurate representation of the interface between the soil and conduit is not likely to have affected significantly the outcome of the analyses presented earlier. The reason for making this inference is that the presence of excessively large interface shear beyond the frictional capacity of the bond would result in large variations of thrust around the conduit; however, by comparison of the analytical and actual thrusts shown in Figs. 3.11 and 3.18, respectively, it can be seen that the variations of thrusts predicted by the analysis are not as large as those observed in practice. This tends to confirm that the interface shears between the soil and conduit, as predicted by the FE analysis, are not so large as to be unrealistic.

3.4   Strength Analysis of Conduit Wall

It has been noted earlier that the conduit wall of a completed soil-steel bridge is predominantly in compression. A flexurally weak conduit wall is able to sustain significantly high thrust because it is laterally supported against buckling by the compacted backfill around the conduit. The backfill provides a continuous and nearly elastic support to the conduit wall; this support condition is usually idealized for strength analysis either as a column supported by discrete springs or as a column with a continuous spring support. The former analogy is shown in Fig. 3.22b.

Despite the fact that in many cases the buckling capacity of the conduit wall governs the design of the conduit wall of a soil-steel bridge, only a limited amount of work has been done to estimate this buckling realistically. For the most part, the work that has been done has been limited to calculation of the initial buckling load, sometimes with the assumption that the supporting medium is capable of sustaining both compression and tension. It is obvious that this assumption is not valid, because the supporting medium cannot transmit any tension to the conduit wall.

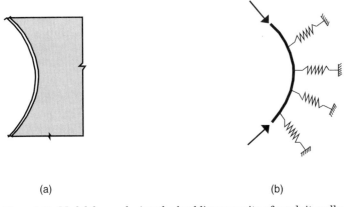

(a)   (b)

Figure 3.22 Model for analyzing the buckling capacity of conduit walls:
(*a*) a segment of the conduit wall and the surrounding soil; (*b*) analogy of
curved column supported by springs.

Many soil-steel bridges are known to be functional even after the
development of local buckling in the conduit wall, thus suggesting that
the wall has substantial postbuckling capacity. The phenomenon of
postbuckling capacity can be understood readily if it is visualized that
in a local buckling condition a part of the conduit wall moves away
from the backfill, leaving a cavity in the soil which immediately arches
around the cavity to disperse the load away from the component in
distress.

It is suggested that considerable research should be conducted to
investigate the phenomenon of buckling of the conduit wall in soil-steel
bridges. This research should be directed toward investigating the
postbuckling behavior of the conduit wall by accounting for the three-
dimensional nature of the structure.

3.5   Concluding Remarks

It has been shown that soil-steel bridges, despite being relatively easy
to construct, are quite complex in their behavior. The load effects in the
conduit wall of the structure are influenced by several factors, including
the long-term settlement of foundations and the arching of soil in both
the longitudinal and transverse directions of the conduit. The backfill,
if composed of well-compacted granular material, can be analyzed by
the assumption of linear elasticity because of its having stress-strain
characteristics that are reasonably time-independent. The foundation
on which the structure rests may have time-dependent characteristics
which should be taken into account when analyzing the effect of foun-

dation settlement on load effects in the conduit wall. It is postulated that some soil-steel bridges suffer distress some time after their construction because of the long-term settlement of the foundation.

For accurate and realistic force and strength analyses of soil-steel bridges, it is necessary that the method of analysis should be able to do the following:

1. Idealize the structure in three dimensions.

2. Represent realistically the properties of both the backfill and the foundation.

3. Deal with the various stages of construction, including placement of layers of backfill on either side of the conduit, first in its loose state and then in the compacted state.

4. Represent the second-order effects resulting from the change of the geometry of the structure.

Another complication which arises in analyzing the behavior of soil-steel bridges is the significant variability of the properties of the soil from point to point in the same backfill. Because of this variability, the behavior of soil-steel bridges cannot always be analyzed by deterministic methods alone. Also because of this variability, it is often very difficult to verify methods of analysis by comparing their results with those obtained by model or field-testing.

This chapter, instead of giving recipes for accurate analysis of soil-steel bridges, has attempted to identify the factors which influence significantly the behavior of these structures. It is hoped that these factors can be taken into account in future analytical studies into the behavior of soil-steel bridges, and that these studies will yield simple-to-use, yet accurate, methods of designing them.

Reference

1. Bakht, B., and Agarwal, A. C., "On Distress in Pipe-Arches," *Canadian Journal of Civil Engineering,* 15(4), 1988, pp. 589–595.

Structural Design Philosophy

George Abdel-Sayed

*University of Windsor, Windsor, Ontario,
Canada*

This chapter deals with the philosophy of designing soil-steel bridges and contains the details of the various design analysis procedures; it should be regarded as a companion chapter to Chap. 5, which contains the specifics of the relevant design analysis procedures required by the various bridge design codes of North America.

In principle, the structural design of a soil-steel bridge is concerned with the design of both the soil and metallic shell; in practice, however, it involves primarily the selection of the metallic shell. This selection, which indeed does implicitly take account of the composite nature of the soil and metallic shell, requires the determination of load effects in the shell and the prediction of its load-carrying capacity. The former process is usually referred to as *force analysis,* and is concerned with the determination of such load effects as thrusts, moments, and deformations of the metallic shell during and after the construction of the structure. The latter process, called *strength analysis,* is concerned with the capacity of the structure to sustain the calculated load effects, it being noted that, as explained further in this chapter, the strength analysis is also related explicitly to the load-carrying capacity of the soil.

As discussed in Chap. 3, the load effects in a soil-steel bridge depend upon various factors, such as the geometry of the structure, dispersion of concentrated loads through the backfill, the material properties of the metallic shell, and those of the surrounding soil, which may or may not be time-dependent. While it is desirable to include all these factors rigorously in the design analysis, this may not be practicable. Accordingly, a balance has to be struck between the accuracy of the calculations and the ease of their application in the design office. The appropriate compromise between accuracy and ease of application depends upon the size and cost of the structure, as well as on the consequences of its failure.

As shown later in this chapter, design analysis of soil-steel bridges ranges between extremely easy-to-apply empirical design methods and fairly complex methods involving the use of finite element analyses.

4.1 Loading

As discussed in Chap. 3, Sec. 3.2, the load effects in the conduit wall and surrounding soil depend not only upon the load themselves, but also upon the material properties of the various components of the structure and the geometry of the conduit. Assumptions made for the various loads in the force analysis are an integral part of the design process. The various loads that are usually considered in the design of a soil-steel bridge are dead loads, live loads (including dynamic load allowance), and earthquake loads. These loads are discussed separately and in some detail in the following subsections. Despite the non-linear response of a soil-steel bridge to loads, the load effects due to various loads are customarily superimposed, it being assumed that errors involved by this practice are negligibly small for the purposes of design.

4.1.1 Dead loads

The soil above and around the conduit and the wearing course on the embankment above the conduit constitute the dead loads in a soil-steel bridge. The weight of the metallic shell itself, being relatively small, is not usually included in the dead-load analysis. The dead-load effects in the metallic shell can be established realistically only by following the history of loading during the various stages of construction. However, for practical reasons, the force analysis of dead-load effects in the metallic shell are related either to the soil pressure at the crown of the conduit, or the weight of the column of soil directly above the conduit. It is noted that for the no-arching condition, the conduit wall thrust T_D due to dead load is exactly equal to $W/2$, where as shown in Fig. 4.1, W is the weight of the column of soil and the wearing course directly above the conduit.

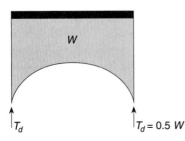

Figure 4.1 Conduit wall thrust in the condition of no arching.

It is noted also that the vertical soil pressure at the crown level varies across the conduit span, as shown in Chap. 3, Fig. 3.10, and that it is affected by the degree of soil arching. As discussed in Chap. 3 and shown in Fig. 3.7*b*, the tendency of the column of soil immediately above the conduit to move downward relative to adjoining soil masses induces positive arching, which helps to relieve the conduit wall of some of its thrust. Conversely, as shown in Fig. 3.7*c*, negative arching is induced when the adjoining soil masses tend to move downward with respect to the column of soil above the conduit. When negative arching takes place, the conduit wall attracts more thrust than that caused by the column of soil above the conduit alone.

The soil in the engineered backfill zone, which is identified in Fig. 1.7 (Chap. 1), is assumed to have a uniform density despite its inherently nonhomogeneous nature. However, in the calculation of the free-field overburden pressure at the crown level, it is customary to take account explicitly of the different densities of the various materials that may exist in a structure. One example of different materials above the conduit is shown in Fig. 4.2.

It is important to remember that the inherent variation of density in a given material is implicitly taken into account in the design by such devices as load factors and safety factors.

4.1.2 Live loads

It is not feasible to design a bridge individually for each of the millions of vehicles that is likely to cross it during its lifetime. Accordingly, bridge design codes specify a limited number of design loadings which are representative of the actual traffic. The design loadings, which comprise concentrated, knife-edge, or uniformly distributed loads, are formulated in such a way that the load effects induced by them in any bridge component constitute, with a known degree of certainty, an

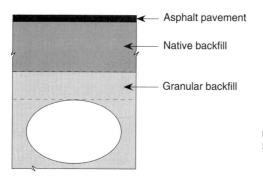

Figure 4.2 Example of different materials above the conduit.

upper bound of the corresponding load effects caused by all actual or foreseen vehicles that are expected to cross the bridge. Almost without exception, the bridge design loadings are formulated on the basis of maximum moments and shears in one-dimensional beams and, in most cases, the beams are assumed to be simply supported.

The design loadings formed on the basis of moments and shears in beams are usually adequate for conventional bridges in which the governing load effects are related to beam moments and shears. As shown conceptually in Fig. 4.3, a vehicle on a simply supported bridge can be represented realistically by a uniformly distributed load. However, such a uniformly distributed load may not be able to represent adequately the vehicle loading on a soil-steel bridge. This is because the concentrated loads of the wheels of a vehicle disperse through the fill in such a manner that the equivalence between the actual vehicle and the idealized design loading, which is valid for beam-type bridges, is no longer maintained for soil-steel bridges.

The maximum load effects in a soil-steel bridge are caused either by single axles of vehicles or by groups of closely spaced axles. Usually, dual-axle groups are found to be more critical in soil-steel bridges than are groups with three or more axles. A design loading can be applicable to soil-steel bridges only if it comprises loads which correspond directly to the single and dual axles of actual vehicles. Fortunately, the design loadings of North American bridge design codes, namely AASHTO (Ref. 6), CSA (Ref. 11) and OHBDC (Ref. 30), have such correspondence. Their design trucks relevant to the design of soil-steel bridges are shown in Fig. 4.4*a*, *c*, *d*, respectively. One-half of the heaviest axles of these design vehicles represents dual tires of actual vehicles. The combined effective contact area of these dual tires on the road surface is taken as 600×250 mm (24×10 in) for the three design vehicles. As shown in Fig. 4.4*b*, the latter dimension is measured along the longitudinal direction of the vehicle.

As discussed in Chap. 3, Sec. 3.2, soil-steel bridges are usually analyzed through plane-strain idealizations which are two-dimensional in nature and are based upon the implicit assumption that both dead and live loads do not vary along the conduit axis. By eliminating one

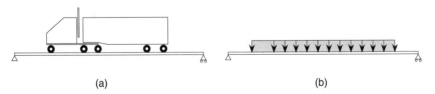
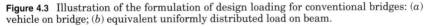

(a)   (b)

Figure 4.3 Illustration of the formulation of design loading for conventional bridges: (*a*) vehicle on bridge; (*b*) equivalent uniformly distributed load on beam.

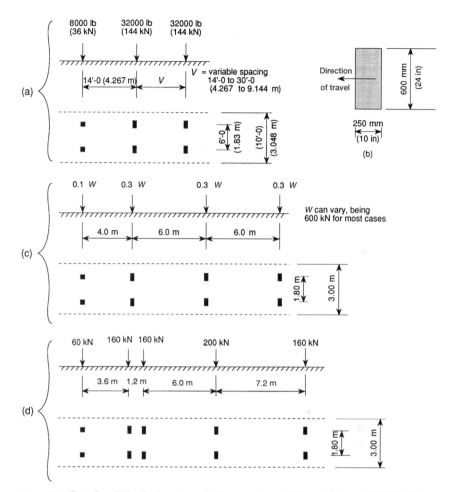

Figure 4.4 Details of North American bridge design vehicles: (*a*) the AASHTO HS-20 vehicle; (*b*) contact area of the heaviest wheel on ground; (*c*) the CSA CS-W vehicle; and (*d*) the OHBDC design truck.

dimension in the idealization, the analysis is made considerably simpler. A two-dimensional plane-strain idealization may be a sound approximation for dead-load analysis, especially when the section under consideration is well away from the ends of the conduit. However, analysis of live-load effects through such an idealization is not automatically reliable because the live loads are, indeed, not continuous along the conduit axis.

In soil-steel bridges with either or both relatively large spans and shallow depth of cover, the live-load effects constitute a fairly large proportion of the total load effects. For such bridges, particular attention needs to be paid to the representation of concentrated live loads for the plane-strain idealization of the bridge.

4.1.3 Earthquake loads

There appears to have been no comprehensive study up to the present of the effects on soil-steel bridges of loads imposed by earthquakes. However, it seems reasonable to conclude that the response of a soil-steel bridge to earthquake loads should not be markedly different from the response of an equivalent embankment without a conduit running through it. Accordingly, it is recommended that the same considerations be given to earthquake loading on soil-steel bridges as those that are applied to the earthquake design of embankments.

A fundamental requirement for the design of embankments in earthquake-prone zones is that the backfill material as well as the engineered material used in the bedding zones should be so selected and placed that it does not liquify during an earthquake. Liquefaction is the phenomenon in which a cohesionless soil loses strength during an earthquake and acquires mobility, permitting fairly large movements. A review provided in Ref. 43 may be found useful in assessing the liquefaction potential of soil at a given site.

It may be noted that a soil-steel bridge which is constructed with soils that do not liquify, and which remains stable during earthquakes, is ideally suited to absorb the energy imparted by earthquakes.

4.2 Design Criteria

Any engineering structure is designed to remain in service with a high degree of certainty for the various functions for which it is intended. In bridges, these functions may, for example, require that the structure not deform excessively under normal, everyday traffic; additionally, it may be required that under exceptional loads, which are assumed to occur only once or twice during the lifetime of the structure, the bridge shall not fail, although it may in certain cases experience large deformations. The former requirement relates to the "serviceability" of the bridge and the latter to its ultimate load-carrying capacity.

Current design procedures for soil-steel bridges do not account for serviceability and are concerned only with the ultimate load-carrying capacity. As discussed in Sec. 4.1, these structures are usually designed only for dead and live loads, with the latter corresponding to vehicle weights. It is noted that if earthquake loading is accounted for in the design of the embankment, there is no need to consider it specifically in the design of the soil-steel bridge.

According to the load factor design (LFD) method, the load and resistance factor design (LRFD) method, and the limit states design (LSD) method, the design criterion involving only the dead- and live-load effects on a component of the structure can be written as

$$\phi R_n \geq \alpha_D L_D + \alpha_L L_L(1 + DLA) \tag{4.1}$$

where ϕ = the resistance, or performance, factor accounting for variations in the strength of the component
R_n = the nominal, or specified, strength of the component
α_D = the dead-load factor accounting for variations in dead-load effects
L_D = the nominal dead-load effect
α_L = the live-load factor accounting for variations in the combined effects of live loads and impact or dynamic load effects
L_L = the nominal live load effect
DLA = the impact factor, or dynamic load allowance

The differences between the three design methods mentioned earlier lie in the manner in which the various factors are calculated. While it may be advantageous for a designer to become familiar with the philosophy and formulation of the design method, it is not essential. Unlike the LFD method, the LRFD and LSD methods are based upon nominal probabilities of failure which are quantified by a safety index usually denoted by β (beta).

The working-stress method of design is not recommended to be used for soil-steel bridges and, accordingly, is not discussed here.

In general, the modes of failure of a soil-steel bridge are interrelated. However, for simplicity it can be assumed that the overall failure is triggered by one of the following local modes of failure

- yielding of conduit walls
- buckling of conduit walls
- failure of the seams of a conduit wall
- soil failure above the conduit
- bearing failure of the soil

The procedures and measures for designing against these potential failures are discussed subsequently under separate headings, along with the procedures to account for the safety of the structure during construction.

4.2.1 Safety during construction

Despite the fact that the conduit walls of soil-steel bridges on their own are very flexible, they need to have a certain flexural stiffness and moment capacity if they are not to deform excessively during construction. Development of plastic hinges during construction, while expected in certain cases, can significantly affect the subsequent behavior of the completed structure.

The behavior of the pipe during the various stages of construction is explained in Ref. 21 and is reproduced in Fig. 4.5; it may be found helpful in understanding the relevant design requirements.

As shown in Fig. 4.5a, during the initial stages of backfilling the active horizontal pressure is resisted only by the bending rigidity of the pipe. When the fill height grows above the spring line, as shown in Fig. 4.5b, the equilibrium of the pipe is maintained not only by the flexural rigidity of the pipe, but also by the vertical component of the dead load of soil and the passive soil pressure in the haunch areas, the latter being developed by the tendency of the pipe in these areas to move toward the soil. When the fill height reaches almost to the crown, the resulting soil pressures are shown schematically in Fig. 4.5c. It can be seen in Fig. 4.5b, c that, during the later stages of backfilling, the upper portions of the pipe move inward. The bending moments and displacements of the pipe are reduced, or even reversed, as the height of the backfill is raised above the crown.

The main design consideration during the backfilling operation is that the conduit wall should not be permanently deformed as a result of the induced stresses exceeding the yield stress of the conduit wall material. It is also necessary that the deformation of the pipe during construction should remain within tolerable limits so that its postconstruction performance is not affected adversely.

As will be discussed in Chap. 6, Sec. 6.5, the pipe deformations during construction can be controlled readily by several means, such as surcharging over the pipe. Notwithstanding such remedial measures, it is desirable to predict by analytical means the potential for deformation problems even before the construction begins. Analytical methods are available, such as the one presented in Sec. 4.3, the use of which will allow pipe deformations during construction to be predicted with fair accuracy. Certain requirements relating to the flexibility of the pipe, which are specified in various codes and discussed in this section, are also helpful in ensuring that the pipe does not deform excessively during construction.

4.2.2 Yielding of conduit walls

Some experts in the field of soil-steel bridges believe that when there is adequate depth of soil cover over the pipe, the failure of the conduit

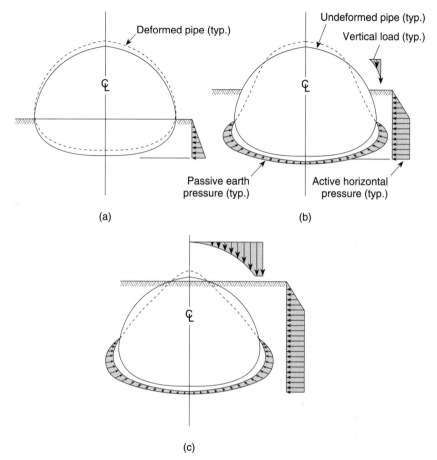

Figure 4.5 Equilibrium of the pipe during various stages of construction: (*a*) fill height below spring line; (*b*) fill height above spring line; (*c*) fill height near crown.

wall takes place through crushing (i.e., yielding of the entire section). They therefore calculate stresses in the conduit wall due only to circumferential thrust, with no regard to stresses due to bending. The justification for neglecting these bending moments lies in the manner in which the interface pressure between the conduit wall and the surrounding soil changes with movement of the wall. According to this line of reasoning, if the bending moment causes local partial yielding of the conduit wall, the resulting movement of the wall will be such as to cause a local increase in the interface pressure, which consequently inhibits the development of further bending moments. It should be emphasized that this neglect of bending moments is justifiable only when the depth of cover is large and the flexural rigidity of the conduit wall relatively small, as it is for the 152 × 51 mm (6 × 2 in) profile commonly used in soil-steel bridges. When the depth of cover is shallow, say less than one-fourth span, the backfill around the pipe may be insufficient to give complete restraint to the conduit wall movement, especially under unsymmetric loads, in which case the induced moments cannot be omitted from design consideration.

Some experts assume that the conduit wall reaches its ultimate limit when the combined stresses due to the axial force and bending moments reach the yield point of the steel at any point on the section, which is usually the crown. In the context of such assumption, it should be recognized that the local yielding of the corrugated plate, or even the formation of a plastic hinge at one location, does not constitute the failure of the statically indeterminate soil-steel bridge. In this structure, failure due to bending effects is expected to occur only after the formation of a sufficiently large number of hinges to form a failure mechanism.

As subsequently discussed, failure by yielding provides only the upper bound of the failure load, which in reality may be more closely related to buckling instability than to pure crushing.

4.2.3 Buckling of conduit walls

Earlier investigators paid little attention to buckling as a potential mode of failure of the conduit wall. The failure of the conduit wall was generally believed to be related either to wall crushing or to deformations of the pipe (Ref. 38). Field performance of pipes having diameters less than about 3 m (10 ft) and fairly large depths of cover has provided some support for this belief.

Deflection control as a means of avoiding buckling was recommended in Ref. 38 after experimental work which suggested that the pipe reaches a state of incipient collapse when its vertical diameter is reduced by 20 percent as a result of overburden pressure. Following this investigation, it became customary to assume that failure of the pipe corresponds to a 20 percent reduction in the rise. Accordingly, a 5 percent reduction in the rise caused by overburden pressure was taken as allowable, implying a "factor of safety" of 4.0 against failure.

In recent years, a general increase in the size of soil-steel bridges has prompted further refinement of the design process, which in turn has led to further research on the phenomenon of failure of the conduit wall. This research has brought to light the fact that buckling is a significant mode of failure, and that it can take place even when the decrease in the conduit rise is less than 5 percent. It is now well established that the stability analysis of the conduit wall is an indispensible component of the design process, especially when the spans are large and depths of cover small, and that a conduit wall can buckle well before it will "crush."

The analysis of the buckling of the conduit wall is made particularly difficult by the geometric nonlinearity of the composite structure as well as the nonlinear behavior of soils and the formation of plastic hinges in the conduit wall. Details of strength analyses dealing with the buckling of conduit walls are given in Sec. 4.5.

4.2.4   Failure of seams

As will be discussed in Chap. 6, Sec. 6.4, the curved, corrugated steel panels are bolted together with 20 mm (¾ in) diameter bolts to form the pipe. The bolt holes are usually 25 mm (1 in) in diameter to facilitate the insertion of bolts and to permit some flexibility during the assembly of the pipe. While the bolts are lightly tightened during initial assembly of the pipe, they are fully torqued before the placement of the backfill. Hence the seams are virtually locked before the backfilling operation starts.

Since the loads on the pipe do not vary significantly along its axis, the circumferential seams are not designed from structural considerations; the spacings of bolts in these seams are fixed by the corrugated-steel-plate industry for reasons other than structural. Typical spacings of bolts in circumferential seams are shown in Fig. 1.18 (Chap. 1). The longitudinal seams, unlike the circumferential seams, have to transmit mainly thrusts from one plate segment to another. Accordingly, bolts in the former seams must be designed to transmit the thrust through shear in them.

There are only three arrangements of bolts for longitudinal seams in pipes constructed of 152×51 mm $(6 \times 2$ in$)$ corrugated plates; these arrangements, incorporating two, three, and four bolts per pitch length of the corrugated plate, are identified in Chap. 1, Fig. 1.18a, b, c, respectively. As shown in Ref. 23, the arrangement with four bolts per pitch length makes the seam too brittle and is not recommended for use in soil-steel bridges. The thrust-transmitting capacities of the other two bolting arrangements are listed in Table 4.1 for different plate thicknesses. These strengths are derived from the test data reported in Ref. 7.

The bolt holes of longitudinal seams can be subject to tearing when the bending deformations of the pipe become excessive; this phenomenon will be discussed in detail in Chap. 10 and App. B.

4.2.5   Soil failure above the conduit

In soil-steel bridges with shallow covers, the soil above the conduit can fail either as shown in Fig. 4.6a or b. The former kind of failure can be referred to as the *sliding wedge failure;* it is initiated by the shear fail-

TABLE 4.1   Ultimate Strength of 152 × 51 mm (6 × 2 in) Corrugated Steel Plates Jointed with 20-mm- (¾ in) Diameter Bolts, and Subjected to Pure Thrust

Metric units			U.S. customary units		
Nominal plate thickness, mm	Strength in kN/m for bolting arrangement		Nominal plate thickness, in	Strength in kips/ft for bolting arrangement	
	1	2		1	2
3	750		0.109	43.0	
4	1100		0.138	62.0	
5	1400		0.168	81.0	
6	1800		0.188	93.0	
7	2050	2550	0.218	112.0	
			0.249	132.0	
			0.280	144.0	180.0

NOTE: Bolting arrangements 1 and 2, which require 2 and 3 bolts per pitch length, respectively, are identified in Fig. 1.18.

ure of soil when the embankment is subjected to loads that are eccentric with respect to the conduit axis. The latter type of failure, which can be referred to as the *tension failure,* is also caused by eccentric loading. The use of heavy construction equipment during construction is known to have caused failure of the soil above the conduit.

Since the structural design is usually concerned only with the load effects in the conduit wall, little attention is paid to the state of stresses in the soil. Failure of soil above the conduit is often accounted for in the design codes by specifying a minimum depth of cover; such criteria are discussed in Sec. 4.6.

4.2.6   Bearing failure of soil

Because of the high degree of flexibility of the conduit wall, the loads imposed on the pipe are resisted mainly by membrane action of the

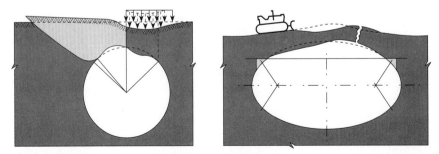

Figure 4.6   Illustration of soil failure above conduit: (*a*) wedge shear failure; (*b*) tension failure.

wall. This phenomenon compels the radial soil pressure on the conduit wall to be approximately inversely proportional to the local radius of curvature of the conduit, as is illustrated in Chap. 3, Fig. 3.5.

Theoretical interface radial soil pressures on pipes of different shapes predicted by the ring compression theory are shown in Fig. 4.7 in solid lines. Consistent with the abrupt changes of the radii of curvature of the actual conduits, the theoretical pressures have sharp discontinuities, as shown in Fig. 4.7*b, c.* Clearly, such discontinuities cannot exist in practice. It is postulated that the actual pressures have

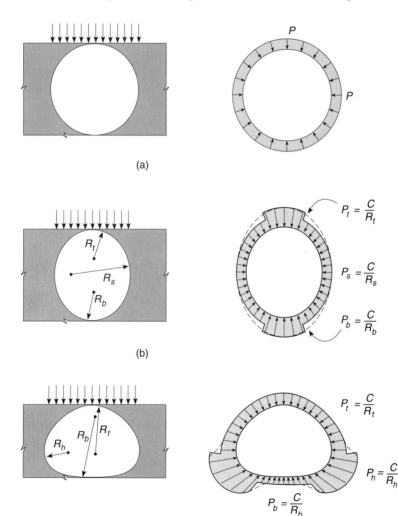

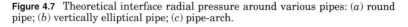

Figure 4.7 Theoretical interface radial pressure around various pipes: (*a*) round pipe; (*b*) vertically elliptical pipe; (*c*) pipe-arch.

smooth transitions from high to low values as also shown by dotted lines in these figures.

Notwithstanding the doubts about the exact patterns of radial pressures, it is clear that soil pressures on conduit walls at locations of relatively small radius of curvature are much higher than those in locations of larger radius of curvature. This is the case, for example, in the haunch areas of a pipe-arch shown in Fig. 4.7c. In such areas of high interface pressure, the soil can fail by deforming excessively. This kind of soil failure is of special concern when the soil in contact with adjoining portions of the pipe that are of larger radius of curvature (because of being subjected to much lower interface radial pressure) does not also undergo excessive deformations. When the pipe is subjected to localized excessive deformations of the soil, it can itself undergo large bending deformations. These, in turn, can lead to significant distress in the pipe or even to the failure of the whole structure, as will be discussed in Chap. 10.

The localized bearing failure of the soil around the pipe can be prevented by limiting the differences in the radii of curvature of plates meeting at a longitudinal seam, and by ensuring that the backfill in areas of locally high-interface pressures is stiff enough. It is noted that the OHBDC (Ref. 30) places limits on the differences in radii of curvature by specifying that the radius of curvature of the conduit wall at any location away from the crown should not be less than one-fifth of the radius at crown. This code also specifies that in pipe-arches the foundation (i.e., the natural ground under the pipe) must be strengthened by digging trenches on either side of the conduit according to the scheme of Fig. 4.8. These trenches are then required to be filled with good-quality and well-graded granular material compacted to at least 95 percent standard Proctor density. Such strengthening is not required when the foundation is of clay and its preconsolidation pressure exceeds the radial pressure at the haunches due to factored loads.

In clay foundations which need strengthening, the depth of the reinforcing trenches is required to be such that the dispersed haunch pressure at the bottom of the trenches does not exceed the preconsolidation pressure of the foundation at the corresponding level.

It should be noted that the strengthening trenches can encourage water seepage through them; for this reason they should be provided with adequate protection against seepage at both the inlet and outlet ends.

4.3   Force Analysis during Construction

As discussed in Sec. 4.2, the metallic shell of the soil-steel structure must possess adequate flexural rigidity to remain undamaged during

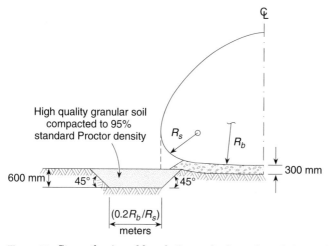

Figure 4.8 Strengthening of foundation under haunches of pipe-arches.

the backfilling operation. The selection of the minimum flexural stiffness of the conduit needed to ensure this can be based either on empirical approaches or on a number of different analytical methods. These empirical approaches and analytical methods are discussed subsequently under separate headings, it being noted that all of the analytical methods presented herein are based on the plane-strain idealizations.

4.3.1 Empirical formulas

An empirical formula was first introduced by the corrugated metal pipe industry to control the flexural rigidity of the pipe for handling during construction. According to this formula, the moment of inertia I of the conduit wall per unit length of the pipe should not be less than $p^2/k\pi^2$, so that

$$I \geq \frac{p^2}{k\pi^2} \qquad (4.2)$$

where I = the conduit wall moment of inertia in (cm^4/m)
  p = the perimeter of the pipe (cm)
  k = an empirically derived factor whose suggested value is 2000

According to a similar approach introduced in Ref. 7, the flexural rigidity of circular pipes is controlled by means of a flexibility factor FF, which is defined by

$$FF = \frac{D^2}{EI} \qquad (4.3)$$

where, for the version in U.S. customary units, D = the diameter, or span, in inches, the modulus of elasticity E of steel is in psi and the moment of inertia I of the conduit wall is in ⁴/in. This approach required that FF should not exceed 0.02 in/lb. It is interesting to note that this restriction coincides with the earlier one for circular pipes if k is taken to be 2362.

The AASHTO specifications (Ref. 6) also require an upper limit of FF as 0.02 in/lb for round and elliptical pipes, and 0.03 in/lb for arches and pipe-arches.

The OHBDC (Ref. 30) uses a similar approach, according to which the value of FF is recommended to be not larger than 0.114 mm/N unless experience justifies the use of a higher value. For this metric version of Eq. (4.3), D is in millimeters, E in MPa and I in mm⁴/mm. It should also be noted that the OHBDC provision differs from the AASHTO one in that its D is defined as the equivalent diameter, which is obtained by dividing the pipe perimeter by π.

All of the four approaches described here give closely similar limits for the minimum flexural rigidity of the conduit that is desirable from the point of view of handling the pipe during construction and its structural integrity during the backfilling operation. It is emphasized that these limits are only empirically derived and that they can be ignored if, with improved methods of construction, it becomes possible to erect a pipe of lesser rigidity.

4.3.2 Duncan method

Reference 12 contains the details of a finite element study by Duncan of the behavior of corrugated metal plate pipes during backfilling and with shallow depths of cover. This study confirmed that the general deformations of the pipe during construction are as shown conceptually in Fig. 4.5. It also confirmed that an increase of backfill stiffness leads to a decrease of bending moments in the conduit wall, while an increase in the flexural stiffness of the conduit wall leads to an increase in them. In conformity with the observations of others (e.g., Ref. 37), Ref. 12 shows that the bending moments in the pipe are characterized by a dimensionless parameter N_f which is defined by

$$N_f = \frac{E_s D_h^3}{EI} \tag{4.4}$$

where E_s = the secant modulus of the backfill, the values of which
  depend upon the quality and compaction of backfill and
  the depth of cover (values of E_s are provided in Chap. 7)
D_h = the conduit span
E = the modulus of elasticity of the material of the conduit wall

I = the moment of inertia of the conduit wall per unit length of the pipe

It is further shown in Ref. 12 that the bending moments per unit length of the conduit wall, denoted as M_B, are given by

$$M_B = R_B(K_{m1} \gamma D_h^3 - K_{m2} \gamma D_h^2 H) \qquad (4.5)$$

where γ is the weight of the backfill per unit volume; D_h is the conduit span; H is the depth of cover; R_B is a dimensionless reduction factor which depends upon the rise to span ratio; and K_{m1} and K_{m2} are dimensionless coefficients which depend upon N_f. K_{m1}, K_{m2}, and R_B are given by

$$\begin{aligned}
K_{m1} &= 0.0046 - 0.0010 \log_{10} N_f & &\text{for } N_f \le 5000 \\
K_{m1} &= 0.0009 & &\text{for } N_f > 5000
\end{aligned} \right\} \qquad (4.6)$$

$$\begin{aligned}
K_{m2} &= 0.018 - 0.004 \log_{10} N_f & &\text{for } N_f \le 5000 \\
K_{m2} &= 0.0032 & &\text{for } N_f > 5000
\end{aligned} \right\} \qquad (4.7)$$

$$\begin{aligned}
R_B &= 0.67 + 0.87(\frac{R}{D_h} - 0.20) & &\text{for } 0.20 \le \frac{R}{D_h} \le 0.35 \\
R_B &= 0.80 + 1.33(\frac{R}{D_h} - 0.35) & &\text{for } 0.35 < \frac{R}{D_h} \le 0.50 \\
R_B &= 2\frac{R}{D_h} & &\text{for } 0.50 < \frac{R}{D_h} \le 0.60
\end{aligned} \right\} \qquad (4.8)$$

In the preceding equations, R is the vertical distance between the crown and the top of the footing in case of nonreentrant arches, and between the crown and spring lines for other structures.

By using Eqs. (4.5) through (4.8), the bending moments in the conduit wall can be calculated for H varying between 0.0 and 0.25 D_h. The maximum thrust T_D in the conduit wall corresponding to the depth of cover within this range can be obtained from

$$T_D = K_{p1} \gamma D_h^2 + K_{p2} \gamma H D_h \qquad (4.9)$$

where K_{p1} is a dimensionless coefficient which accounts for backfill up to level of the crown, and K_{p2} is a dimensionless coefficient which accounts for backfill above the crown; these coefficients are given by

$$K_{p1} = 0.20\frac{R}{H} \qquad (4.10)$$

$$K_{p2} = 0.90 - 0.50\frac{R}{D_h} \qquad (4.11)$$

R is again as defined previously.

It is suggested in Ref. 12 that the combination of moment and thrust should be such that the factor of safety corresponding to the formation of the plastic hinge under these responses is a minimum of 1.65.

Reference 12 also provides the following expression for estimating the crown deflection δ_{crown} when the backfill is at the level of the crown.

$$\delta_{\text{crown}} = \frac{K_\Delta \gamma D_h^5}{100,000EI} \tag{4.12}$$

where δ_{crown} is in inches and K_Δ is a dimensionless coefficient which depends upon N_f and whose values are reproduced in Fig. 4.9.

It is noted that the finite element analyses on which the values of K_Δ given in Fig. 4.9 are based were derived on the assumption that the behavior of the structure is elastic and that the backfill is placed symmetrically on both sides of the conduit. As will be discussed in Chap. 6,

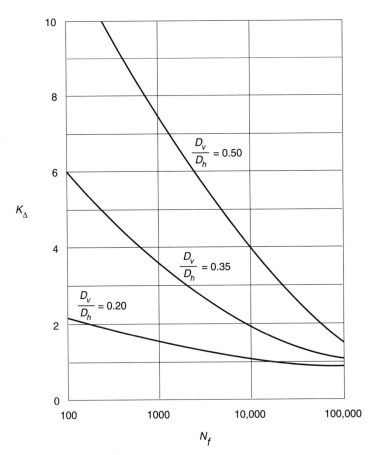

Figure 4.9 Chart for K_Δ.

unsymmetrical backfilling and the presence of heavy equipment can deform the pipe substantially, causing a rolling type of deformation and even failure.

4.3.3 Ontario method

As was noted earlier, the OHBDC (Ref. 30) recommends that the flexibility factor *FF* should not exceed 0.114 mm/N under normal circumstances. In addition, this code controls the flexibility of the pipe by specifying that the upward crown movement of circular and elliptical pipes during backfilling should not exceed δ_{crown} which is given by

$$\delta_{crown} = 1000 \; \mu \; \frac{D_h^2}{d} \qquad (4.13)$$

where δ_{crown} is the vertical movement of the crown in meters with respect to the invert elevation; D_h is the conduit span in meters; d is the depth of the corrugation profile in millimeters; and μ is a dimensionless coefficient which depends upon the ratio D_h/D_v and whose values are reproduced in Fig. 4.10. For other conduit shapes, the OHBDC requires that the upward crown movement during backfilling should not exceed 5 percent of the rise of the conduit.

The controlling limits of the crown movement quantified by μ were derived by analyses in which pipes made out of 152×51 mm (6×2 in) corrugation profile steel plates were subjected to uniform lateral pressures as shown in Fig. 4.11. Limits of μ which could be characterized by the ratio D_h/D_v, were established such that the stresses anywhere in

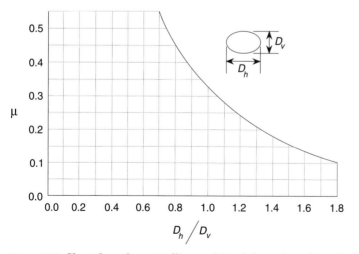

Figure 4.10 Chart for μ for controlling peaking deformation of round and elliptical pipes during construction.

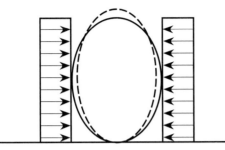

Figure 4.11 Pipe subjected to uniform lateral pressures.

the cross section of the pipe remained within the nominal yield stress of 230 MPa for the steel of the conduit wall.

The method was evaluated in Ref. 25 and was found to correspond to a factor of safety of 1.00. It is recommended in this reference that the factor of safety should be increased to 1.05 in order to account for the fact that perfect symmetry of backfill on either side of the conduit, which was implicit in the derivation of μ, cannot be maintained in practice. The crown movement corresponding to the larger factor of safety is given by

$$\delta_{\text{crown}} = \frac{1000}{1.05}\,\mu\,\frac{D_h^2}{d} \tag{4.14}$$

4.3.4 Frame on elastic supports

The preceding expressions are either empirical in nature or derived approximately from the results of finite element analyses. Further, these expressions are valid for circular or elliptical pipes without circumferential stiffeners or other special features. Rigorous analysis of other cases can be performed readily by using the analogy of frames on elastic supports. This analogy is illustrated in Fig. 4.12 for a segment of the pipe. For further details of this method, reference should be made to Sec. 4.4 in which this analogy is discussed with reference to the analysis of the completed structure.

4.3.5 Summary and conclusions

Despite their apparent differences, all of the empirical and analytical methods given in Sec. 4.3 give similar kinds of results, with their accuracy improving with the rigor of the method. These methods will be found useful in minimizing the risk of problems arising in construction due to the flexibility of the pipe. However, it must be realized that compliance with the constraints implied by these methods does not guarantee that such problems will not arise. Similarly, noncompliance with these constraints does not rule out the possibility that the difficulties

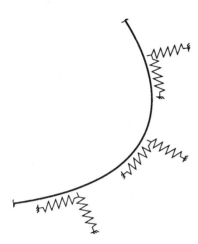

Figure 4.12 Representation of a soil-steel bridge during construction by frame on elastic supports.

arising out of the flexibility of the pipe will not be overcome by improved construction techniques.

It is emphasized once again that the methods given in Sec. 4.3 should be used only as a guide and not as a means of absolute control. It should also be noted that the controls presented in this section relate only to pipes without special features such as transverse stiffeners which are discussed in Chap. 8.

4.4 Force Analysis of Completed Structures

The methods of analyzing load effects in the conduit wall of a soil-steel bridge range in complexity from the semiempirical to those based upon finite element analysis. Some of the commonly used methods are described in this section under separate headings, it being noted that the first two of these methods, i.e., the Marston-Spangler and ring compression methods, are the semiempirical ones. Despite the fact that these methods are no longer commonly used, they are included in the discussion because of their historical significance.

4.4.1 Marston-Spangler method

The Marston-Spangler method described in Refs. 39 and 40 was developed basically for structures with small-diameter round conduits; it is based on the assumption that the top and bottom segments of the pipe of the completed structure are subjected to uniformly distributed vertical soil pressures, and the side segments of the pipe to parabolically distributed horizontal pressures. These soil pressures are illustrated in Fig. 4.13, which shows that the demarcation between the "top" and "side" segments of the pipe is defined by points on the conduit periphery which subtend an angle of 80° at the center of the conduit.

The intensity of horizontal soil pressure on the pipe is assumed to be proportional to the horizontal movement of the pipe. Further, it is assumed that the ratio of the maximum horizontal to vertical pressure depends upon the soil stiffness. When the backfill consists of a well-compacted soil, providing uniform support to the pipe, the maximum horizontal pressure on the side segments is up to 35 percent greater than the vertical pressure on the top segment, as shown in Fig. 4.13.

Based on the preceding assumptions regarding the distribution of soil pressures on the pipe, the thrust in the conduit wall varies from $0.7 P_c R$ at the crown and invert to $1.1 P_c R$ at the haunches where, as shown in Fig. 4.13, P_c is the intensity of vertical pressure on the top segment and R is the radius of curvature of the circular conduit. The corresponding moment in the conduit wall varies from $0.02 P_c R^2$ at the crown, invert and spring lines to about $-0.02 P_c R^2$ at the haunches.

It may be noted that the Marston-Spangler method was a significant step in the development of the analysis of soil-steel bridges. However, this theory has since been replaced by theories which can account more realistically for the behavior of these structures.

4.4.2   Ring compression method

It was proposed in Ref. 46 that the nonuniformity of soil pressure on the pipe assumed in the Marston-Spangler method has little effect on

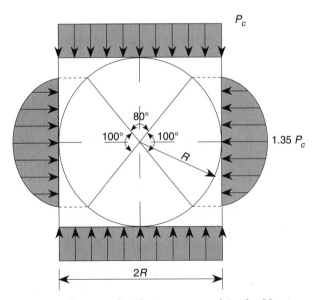

Figure 4.13 Pressure distributions assumed in the Marston-Spangler theory.

the magnitude and distribution of the conduit wall thrust, and that the pipe can be analyzed as a ring in uniform compression. According to its authors, the ring compression method is believed to be valid when the depth of cover exceeds about one-eighth of the conduit span. In this theory, the uniform radial pressure P_c around the pipe is taken as the free-field overburden pressure at the crown level plus the equivalent live-load pressure at the crown level:

$$P_c = \sigma_{V0} + P_L \qquad (4.15)$$

where σ_{V0} is the free-field overburden pressure at the crown level which would have existed in the embankment in the absence of the conduit, and P_L is the equivalent uniformly distributed pressure at the crown level due to live loading including an allowance for impact.

It is assumed that the pipe sustains the pressure at the crown level in pure compression so that the thrust T in a round pipe is given by

$$T = \frac{P_c D_h}{2} \qquad (4.16)$$

In an arch structure having the conduit wall of constant radius of curvature, the thrust in the conduit wall is given by

$$T = \frac{P_c D_h}{2 \cos \theta} \qquad (4.17)$$

where θ defines the position of the junction of the conduit wall with a footing, as shown in Fig. 4.14.

In the case of a noncircular conduit, if the thrust is to have a constant value around the pipe, the radial pressures at the interface of the soil and conduit wall must be inversely proportional to the radius of curva-

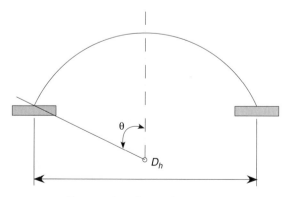

Figure 4.14 Cross section of an arch.

ture of the conduit. While other aspects of the ring compression theory have been open to question, this phenomenon is found to be approximately valid in most soil-steel bridges in which the flexural rigidity of the conduit wall is usually very small in comparison to its axial rigidity. The patterns of radial soil pressures around pipes of different shapes, as assumed in the ring compression theory, are shown in Fig. 4.7.

The ring compression theory has been adopted with some empirical modifications by a few codes and design handbooks. For example, the AISI handbook (Ref. 7) has modified the ring compression theory through an arching factor K which is intended to account for the positive arching that may occur in certain circumstances. According to the AISI method, the conduit wall thrust T is given by the following equation for round pipes

$$T = \frac{K P_c D_h}{2} \tag{4.18}$$

The factor K is taken as unity when the depth of cover is smaller than the pipe diameter. For larger depths of cover, K is assumed to vary with the degree of compaction; its values are taken as 1.00, 0.86, and 0.65 for soils compacted to 80, 85 and 95 percent standard Proctor densities, respectively.

The ring compression theory in the form just discussed proved to be a useful tool in the earlier stages of the development of soil-steel bridges. It, too, has since been replaced by more rigorous and general techniques which are discussed subsequently.

4.4.3   Pipe in an infinite elastic half-space

There are several solutions available in the technical literature for the problem of a pipe buried in an elastic medium; one such solution is presented in Ref. 10 for a circular pipe buried deeply in a weightless, homogeneous, isotropic linear elastic half-space which represents the soil. The pipe response is determined for loads applied to the plane boundary of the half-space, as illustrated in Fig. 4.15. The solution under consideration, which is reported in Refs. 19 and 22, can be applied manually, and can either neglect or take into account the bond between the soil and the conduit wall.

The limitations of this method include the following:

1. The pipe has to be circular.

2. The pipe has to be buried deeply, corresponding to a depth of cover of more than, say, 1.5 times the pipe diameter.

3. The soil is considered to be weightless and its weight is applied as overburden pressure.

4. The soil is assumed to be linear elastic.

Overburden pressure

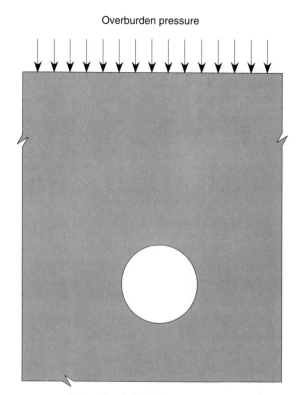

Figure 4.15 A pipe deeply buried in an elastic half-space.

Because of these limitations, this method is only of academic interest and is of little use in the design office.

4.4.4 Plane-frame on elastic supports

It has been proposed in Refs. 21 and 41 that a unit length of the pipe of a soil-steel bridge can be analyzed by idealizing it as a plane-frame supported on springs which represent the backfill and are assumed to transmit compressive and shear forces. A significant aspect of analysis by this method is the conceptual division of the radial earth pressures around the pipe into active and passive pressures, which are illustrated in Fig. 4.16. The active pressure, which is assumed to act on the top portion of the pipe, is in the shape of a half sine wave with its amplitude P_c at the crown. As discussed in the section dealing with the ring compression theory, concentrated live loads acting on the embankment above the conduit are transformed into equivalent uniformly distributed pressure P_L acting at the crown level. When P_L is less than the free-field overburden pressure σ_{V0}, P_c is taken as the sum of P_L and σ_{V0},

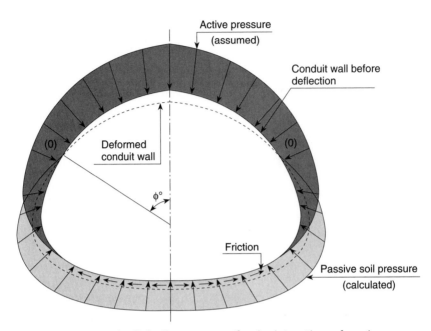

Figure 4.16   Division of radial soil pressures on the pipe into active and passive pressures.

i.e., P_c is given by Eq. (4.15). However, when P_L is greater than σ_{V0}, P_c is given by

$$P_c = 1.10(\sigma_{V0} + P_L) \tag{4.19}$$

The 10 percent increase in P_c reflected by the preceding equation is intended to account for the effect of concentrated loads in the case of very shallow covers. In practice, P_L rarely exceeds σ_{V0}, and therefore Eq. (4.19) is hardly ever used.

The intensity of active radial pressure, p_θ, around the pipe is given by

$$p_\theta = P_c \cos \frac{\pi\theta}{\psi} \tag{4.20}$$

where the notation is as illustrated in Fig. 4.17, and $\psi = \pi$ for shallow covers and $= 1.5\pi$ for deep covers. Active radial pressures in the former case are shown in Fig. 4.17a, in which it can be seen that $p_\theta = 0$ at $\theta = 90°$; the pressures in the latter case are shown in Fig. 4.17b in which $p_\theta = 0.5P_c$ at $\theta = 90°$ and $= 0.0$ at $\theta = 135°$.

While the active earth pressure is accounted for in the analysis through the applied pressure p_θ as just defined, the effects of passive

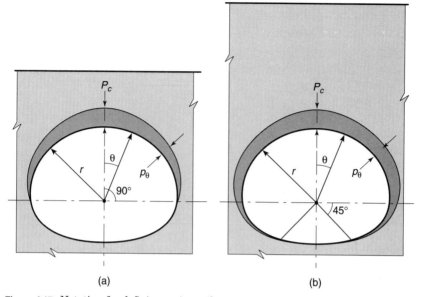

Figure 4.17 Notation for defining active soil pressure: (*a*) structure under shallow fill; (*b*) structure under deep fill.

pressure are taken into account through the springs which represent the backfill. Figure 4.18 shows the applied active pressure on the idealized frame in which the soil in the zone of passive pressures is represented by springs and in which only half the structure is represented, taking advantage of the symmetry of the cross section and loading. It can be seen in this figure that the springs are provided only in the lower portion of the pipe, where passive earth pressures are expected. The exact demarcation between the portions of the pipe which move toward and away from the backfill is not known beforehand. Accordingly, it becomes necessary to use an iterative process in which those springs which are in tension are successively eliminated, with the final iteration corresponding to the case in which all of the springs are in compression.

In the plane-frame idealization, the friction between the conduit wall and the soil can be accounted for by making the springs act at an angle $\tan^{-1} \mu$ to the radial direction, where μ is the angle of friction between the conduit wall and soil.

The accuracy of the method under discussion depends greatly on the accurate representation of the soil stiffness by the springs. The relevant property of the soil which corresponds directly to the stiffnesses of the springs is the coefficient of soil reaction, k_n, which will be discussed in detail in Chap. 7. In this chapter it is shown that the coefficient of soil reaction is not a unique property of the soil and that, in addition to

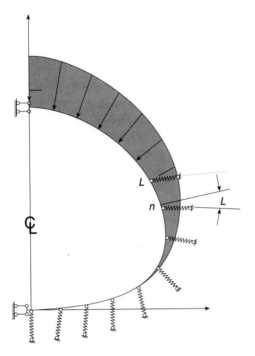

Figure 4.18 Plane-frame on elastic supports.

the elastic properties of the soil medium, it depends upon the size of the conduit, the depth of cover, the location of the reference point and the direction of the idealized spring.

4.4.5   Finite element method

A realistic force analysis of the soil-steel bridge is made difficult by several factors including: (1) geometric nonlinearity of the structure; (2) material nonlinearity with respect to load as well as time; and (3) the dependence of load effects on the history of load application. All of these factors can be taken into account by the powerful and versatile finite element method.

The finite element method solves the mathematical model of a structure by idealizing it as an assembly of components, or elements, which are connected to each other at discrete points, or nodes. The degrees of freedom of displacement at each node are predecided, and their load-deformation characteristics evaluated. By using the various elements, even the most complex structures can be idealized and hence analyzed by the finite element method. It is noted, however, that this method cannot be used as a design office tool by those who are not already familiar with its mechanics and use. The finite element method is too complex to be described in detail in a brief subsection of this book.

Accordingly, this subsection contains only a general description of its use for the analysis of soil-steel bridges. For further reading on the subject, the reader is directed to Refs. 17, 19, and 20 which deal specifically with the analysis of soil-steel bridges by the finite element method.

A soil-steel bridge can be analyzed by the finite element method either through two- or three-dimensional idealizations. In practice, the latter idealization is rarely used, mainly because of the very large amount of computing power associated with it. It is much more common to use the two-dimensional idealization, in which a transverse slice of unit width is assumed to be in a state of plane-strain. As discussed in Chap. 3, Sec. 3.2, in a plane-strain idealization, deformations perpendicular to the plane of the idealized structure are assumed to be zero. The advantage of this idealization is that, by eliminating one dimension from consideration, the computing power required to perform the analysis is reduced considerably. The main disadvantage, as discussed in Sec. 3.3 of Chap. 3 is that the analysis cannot account for three-dimensional effects such as unequal foundation settlement along the conduit length.

A typical mesh representing the plane-strain slice of a soil-steel bridge with a round conduit is shown in Fig. 4.19. It can be seen in this figure that the mesh represents not only the soil envelope on either side of the conduit but also a portion of the foundation below the engineered structure. The mesh can be composed of three types of elements which are described as follows:

Conduit wall elements. The conduit wall is idealized by one-dimensional beam or beam-column-type elements. The former type may suffice if the analysis is linear. However, the latter type of elements are necessary for nonlinear analysis.

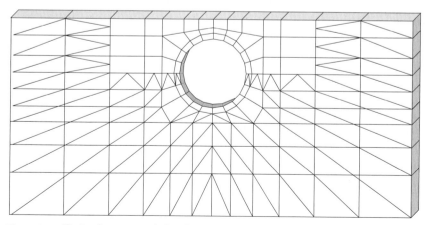

Figure 4.19 Finite element mesh for plane-strain analysis.

Soil elements.   The backfill and foundation are represented by either or both of triangular and quadrilateral elements. In terms of complexity, these elements can range from the simple, constant strain elements representing linear elastic behavior to rigorous linear-strain elements representing the nonlinear behavior of the soil. It is noted that the Duncan model (Refs. 12 and 13) is used to represent the nonlinear behavior of granular soil in many of the generally available finite element programs that have been developed specifically for the analysis of soil-steel bridges.

Interface elements.   The interaction between the soil and conduit wall at their interface is accounted for by interface elements; these elements can be of either the line type or spring type. The latter type is shown conceptually in Fig. 4.20. The interface elements can be used both to develop frictional forces or to represent frictionless behavior.

The finite element method can represent the sequence of placing and compacting the fill layer by layer. When used in this way, a prescribed surface load is imposed on the most recently added layer of the backfill. The magnitude of the imposed surface load depends upon the type and weight of the compaction equipment. Incremental load effects resulting from these surface loads are added to the load effects corresponding to the previous load lifts, and the stress-dependent properties for soil and interface elements are updated.

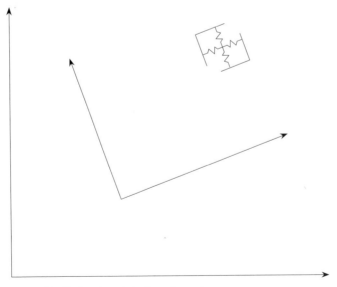

Figure 4.20   Spring-type interface element.

The representation of the compaction process is completed by removing the superimposed surface loads. This removal of the surface loads corresponds to the removal of the compaction equipment. Although the surface loads are removed, the compaction effect of densifying the soil is locked into the elements. This retention involves both the properties and geometry of the elements that take place prior to the removal of the surface loads.

The actual number of lifts, or layers, employed in the placement of backfill in a typical soil-steel bridge is too large to be used conveniently in the finite element analysis. It is usual in such analyses to use a much smaller number of lifts in the simulation of the construction process. For example, the automatic data generation option of the finite element program described in Ref. 20 can generate the mesh corresponding to a maximum of five layers; these layers are identified in Fig. 4.21, in which it can be seen that the bottom layer comprises the foundation of the structure.

The finite element method, perhaps because of its versatility, can easily give wrong answers without the analyst realizing the fact. It is important, therefore, that the use of this method should be left to those who are well-versed with it, and who can readily identify errors in the results, if present.

It is noted that Sec. 3.3 (Chap. 3) contains a discussion of the assumptions employed in the common finite element analysis.

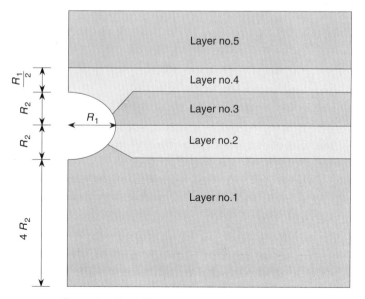

Figure 4.21   Example of backfill layers used in finite element analysis.

Several computer programs (e.g., Ref. 20), incorporating the FE method have been developed specifically for the analysis of soil-steel bridges and other similar buried structures. It is also worth noting that several simplified methods of analysis, some of which are reported in this book, have also been developed with the help of FE methods of analysis.

4.4.6 Ontario method

The Ontario Highway Bridge Design Code, OHBDC (Ref. 30) permits the use of the simplified method given in this subsection for calculating the combined thrust T due to factored dead and live loads. As is customary in the analysis of soil-steel bridges, the effects of dead and live loads are superimposed despite the slightly nonlinear response of the structure under loads. According to the OHBDC method, the conduit wall thrust T due to factored loads is given by

$$T = \alpha_D T_D + \alpha_L T_L \tag{4.21}$$

where T_D and T_L are thrusts due to nominal, or design, dead and live loads, respectively, with the latter including an allowance for impact, and α_D and α_L are dead- and live-load factors which will be discussed in Chap. 5.

The dead load thrust T_D is given by

$$T_D = 0.5(1.0 - 0.1C_s)A_f W \tag{4.22}$$

where W is the weight of the portion of the fill above the conduit as defined in Fig. 4.1; A_f is a coefficient which can be read from Fig. 4.22 and C_s, a dimensionless parameter for the axial rigidity of the conduit, is given by

$$C_s = \frac{E_s^* D_v}{EA} \tag{4.23}$$

where E is the modulus of elasticity of the conduit wall material, E_s^* is the effective secant modulus of soil, and A is the cross-section area per unit length of the conduit wall. It is noted that the values of the effective secant modulus used in conjunction with Fig. (4.23) are given in Chap. 5, Table 5.1. This secant modulus is somewhat different than the secant modulus used in conjunction with other methods. Explanation of the differences between the two are discussed in Chap. 7.

The coefficient A_f depends upon the ratios H/D_h and D_h/D_v, where H is the depth of soil cover above the crown and D_h and D_v are the dimensions of the conduit as defined in Fig. 4.23.

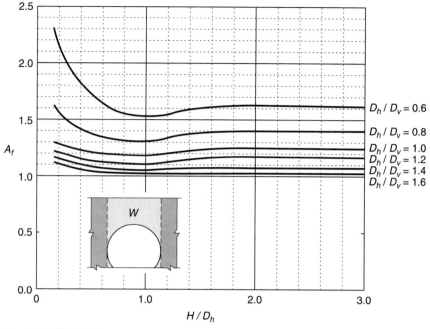

Figure 4.22 Plots of A_f.

The live-load thrust T_L is given by

$$T_L = 0.5\sigma_L m_f(1 + DLA) \times (\text{lesser of } D_h \text{ and } l_t) \qquad (4.24)$$

where σ_L is the equivalent uniformly distributed pressure at the crown level, which is obtained by assuming that the wheel loads at the embankment level are distributed through the fill at an angle of 45° in the transverse direction of the conduit (i.e., its span direction) and at a slope of one horizontally to two vertically in the longitudinal direction; m_f is the multiple presence reduction factor which is equal to 1.0 if only one vehicle is considered on the embankment and 0.9 if there are two vehicles placed side by side; DLA is the dynamic load allowance; and l_t is the length of the dispersed load at the crown level, measured along the conduit span as shown in Fig. 4.24. Chapter 5 will provide the specifics of the OHBDC method, including charts from which the value of $0.5\sigma_L m_f(1 + DLA)$ can be read directly for different depths of cover.

Equation (4.24) is based upon observations from field testing (Ref. 8) and upon conclusions drawn from analytical studies (Ref. 5).

Equation (4.22) is derived from the results of the analytical study reported in Ref. 19. This study shows that the soil immediately above

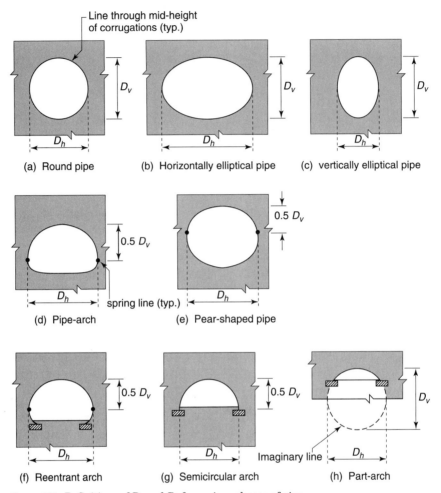

Figure 4.23 Definitions of D_h and D_v for various shapes of pipe.

the conduit is distributed to the conduit wall differently than the soil further above; it also shows that the degree of arching, to which T_D is related directly, depends upon various factors. These factors are listed below, along with a discussion as to how they are accounted for in the specified method. It is noted that the analytical study was conducted through plane-strain idealization of the structure.

Different arching factors.   The fact that portions of the soil above the conduit may have different arching factors can be accounted for by dividing the soil into two portions, namely W_1 and W_2 as identified in Fig. 4.25, and by providing different arching factors for them (i.e., A_{f1} and A_{f2}, respectively).

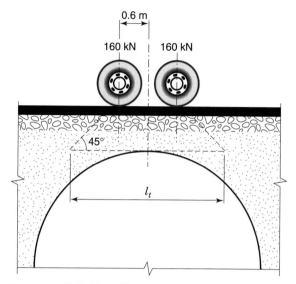

Figure 4.24 Definition of l_t.

Span-to-rise ratio, D_h/D_v. This ratio can also be dealt with explicitly in the charts for A_{f1} and A_{f2} given in Fig. 4.26 *a, b.*

Depth of cover-to-span ratio, H/D_h. This ratio does not apply to A_{f1}. However, for A_{f2}, it can be accounted for explicitly in the charts provided in Fig. 4.26*b.*

Quality of backfill. The quality of the backfill is accounted for in the specified method by restricting its application to structures with backfill consisting of well- to poorly graded granular soil that is compacted to a minimum of 90 percent standard Proctor density.

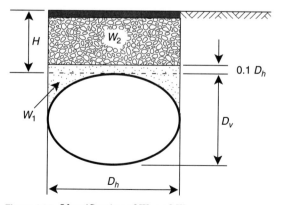

Figure 4.25 Identification of W_1 and W_2.

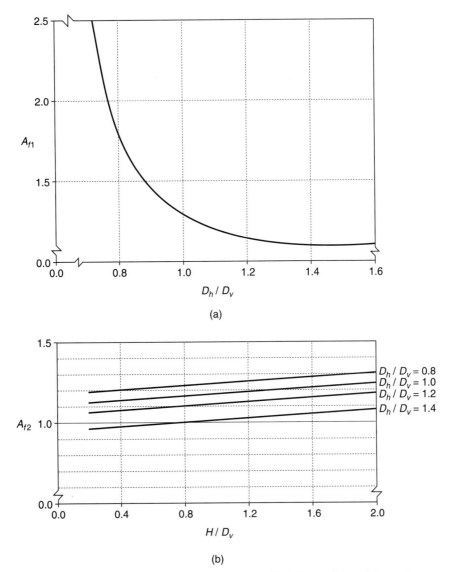

Figure 4.26 Plots of arching factors for structures in which the conduit wall is continuous around the conduit. (a) plot of A_{f1}; (b) plots of A_{f2}.

Width of backfill. The specified method is applicable to structures in which the backfill extends to a minimum of $0.5D_h$ transversely on either side of the conduit, as shown in Fig. 4.27. When the backfill extends farther, the method is expected to give slightly conservative results. It is recalled that the term *backfill* is used for the engineered portion of the total fill.

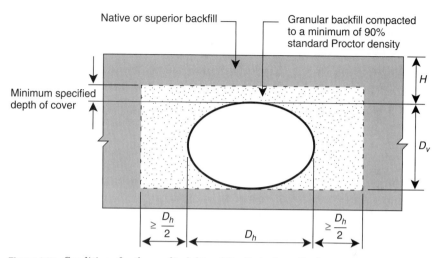

Figure 4.27 Conditions for the applicability of the Ontario method.

Extent of backfill above the conduit. As for the case of backfill width, the specified method is applicable only if the backfill extends above the conduit up to at least the minimum extent specified by the code, as shown in Fig. 4.27.

Foundation. Foundation stiffness affects T_D differently in structures of different conduit shapes. For example, in a structure with a horizontally elliptical conduit, a softer foundation will increase the dead-load thrust in the conduit wall. Conversely, in a structure with a vertically elliptical pipe, the softer foundation will tend to reduce T_D. The values of the arching factors given in Fig. 4.26a, b were selected in such a way that they correspond to foundations which lead to higher values of the thrust.

Conduit wall bending stiffness. The bending stiffness of the conduit wall relative to the stiffness of the backfill is characterized by a dimensionless parameter B_s defined by

$$B_s = \frac{E_s^* D_h^3}{EI} \tag{4.25}$$

It was found that, within the range of the values of B_s encountered in practice, this parameter has little influence on T_D. Accordingly, it was omitted from consideration.

Conduit wall compression stiffness. The compression stiffness of the conduit wall relative to the stiffness of the backfill is characterized by the dimensionless parameter C_s. It was found that this parameter has the effect of reducing A_{f1} and A_{f2} differently; these reductions, denoted as R_{c1} and R_{c2}, respectively, are related approximately to C_s

according to the following equations for the ranges of parameters considered in the analysis, or likely to be encountered in practice.

$$R_{c1} = -0.02 + 0.095C_s \tag{4.26}$$

$$R_{c2} = 0.11C_s \tag{4.27}$$

By including these two reduction factors, the equation for T_D becomes

$$T_D = (1 + 0.02 - 0.095C_s)A_{f1}\frac{W_1}{2} + (1 - 0.11C_s)A_{f2}\frac{W_2}{2} \tag{4.28}$$

It was found that both the multipliers to C_s can be changed to 0.10 without significant loss of accuracy, so that Eq. (4.28) can be rewritten as Eq. (4.22), in which

$$W = W_1 + W_2 \tag{4.29}$$

4.4.7 Duncan method

The same Duncan method which is described in Sec. 4.3 for analysis of the pipe during construction can also be used, after minor modifications, to take account of the effects of live loads. The thrust T in the conduit wall, being the sum of dead-load thrust T_D and live-load thrust T_L, is given by

$$T = K_{p1}\gamma D_h^2 + K_{p2}\gamma H D_h + K_{p3}LL \tag{4.30}$$

where K_{p1} and K_{p2} are the dimensionless factors corresponding to dead loads obtained by Eqs. (4.10) and (4.11), respectively; K_{p3}, the dimensionless factor corresponding to live loads, is given by Eq. (4.31); and LL is the equivalent line load which is obtained as described later in this subsection.

$$K_{p3} = \quad 1.00 \quad \text{for } \frac{H}{D_h} \leq 0.25$$

$$K_{p3} = 1.25 - \frac{H}{D_h} \quad \text{for } 0.25 < \frac{H}{D_h} < 0.75 \tag{4.31}$$

$$K_{p3} = \quad 0.50 \quad \text{for } \frac{H}{D_h} \geq 0.75$$

Ref. 12 shows that the set of concentrated wheel loads on the embankment above the conduit can be represented by a continuous line load of intensity LL by seeking equivalence of peak vertical stresses induced by these loads at the crown level. It is noted that the peak stresses are obtained by Boussinesq elastic theory as described in Ref. 36. The values of LL are given in Table 4.2 for the AASHTO HS-20 vehicle.

TABLE 4.2   Equivalent Line Loads for HS-20 Vehicle
According to Ref. 12

Depth of cover, H, (m) ft	Line load LL	
	in kN/m	in kips/ft
(0.3) 1	88.9	6.1
(0.91) 3	52.5	3.6
(1.52) 5	37.9	2.6
(2.13) 7	35.0	2.4
(3.05) 10	29.2	2.0
(6.1) 20	19.0	1.3

The total moment M in the conduit wall due to dead and live loads is given by

$$M = R_B(K_{m1}\gamma D_h^3 - K_{m2}\gamma D_h^2 H) + R_L K_{m3} D_h LL \qquad (4.32)$$

where K_{m1} and K_{m2} are obtained from Eqs. (4.6) and (4.7), respectively; LL is obtained from Table 4.2 for HS-20 loading; and R_L and K_{m3} are obtained by the following equations.

$$R_L = \frac{3.77 - 0.75 \log_{10} N_f}{\left(\dfrac{H}{D_h}\right)^{0.75}} \le 1.0 \qquad (4.33)$$

$$\left. \begin{array}{ll} K_{m3} = 0.12 - 0.018 \log_{10} N_f & \text{for } N_f \le 100{,}000 \\ K_{m3} = 0.030 & \text{for } N_f > 100{,}000 \end{array} \right\} \qquad (4.34)$$

It is recalled that N_f is the dimensionless parameter corresponding to the flexural rigidity of the conduit wall and that it is defined by Eq. (4.4).

Ref. 12, which gives the details of the Duncan method, recommends that it is not necessary to account for the impact factor in the calculation of thrusts and moments in the conduit wall due to live loads.

Reference 12 defines the demarcation between shallow and deep covers as taken at the depth of cover H being $0.25D_h$, and notes that the bending moments in the conduit wall due to live loads are significant for structures with shallow depths of cover. In such cases, loads placed eccentrically with respect to the crown induce maximum moments in the haunch area. In structures with deep covers, the conduit wall is restrained effectively enough to keep it from deforming excessively under the effects of live loads and thus developing substantial bending moments.

4.4.8   Summary and conclusions

The methods of force analysis given in Sec. 4.4 fall into three categories according to their applicability in the design of soil-steel bridges.

The first category is of methods which are of historical importance but which are no longer recommended to be used in the design office; these methods include the Marston-Spangler and ring compression methods and the method in which the conduit wall is idealized as a pipe in an elastic space of infinite extent. The second category comprises computer-based methods which are recommended for use only as research tools and not for design analysis of soil-steel bridges, save for exceptional cases. This category includes plane-frame and finite element methods. The third category includes the Ontario and Duncan methods which are expected to be of immediate use for designers because they can be applied manually and with minimum calculations.

The Ontario method, because of explicitly or implicitly accounting for all the factors which influence load effects in the conduit walls, is more versatile than the Duncan method; however, it does not give the bending moments in the conduit wall, and these may be of significance if the corrugations are deeper than the usually employed 152×51 mm $(6 \times 2$ in) profile. It is found that for most soil-steel bridges, the Ontario and Duncan methods give similar results.

It is noted that Ref. 19 provides a more general method than the Ontario method, and that, through numerous charts, it can be used to analyze many structures encountered in practice.

4.5   Strength Analysis of Conduit Wall

As discussed in Sec. 4.2, the conduit walls of a soil-steel bridge are subjected to mainly compressive forces, because of which their mode of failure is usually that of crushing or buckling instability. It is usual to design the conduit wall only for withstanding the compressive forces, without paying any regard to bending moments. Such an approach is partially justifiable on the ground that excessive moments can lead to the development of plastic hinges in the conduit wall, and that these hinges are not always detrimental to the structural integrity of the highly indeterminate soil-steel bridge. The effect of bending moments and the development of plastic hinges can, however, be taken into account in the buckling analysis, as shown later in this section.

Earlier investigators paid little attention to buckling as a possible mode of failure of the conduit wall. It was believed that the safety of structures with good-quality backfill could be ensured through control of either the crown deflection or the ring stress. The validity of this approach has been justified up to a point by the in-service performance of structures with circular conduits of less then 3 m (10 ft) diameter.

However, this approach of strength through only deflection control cannot be relied upon for larger structures; in these, as is shown by modern research, buckling of the conduit wall can take place even when the deformations of the pipe are relatively small.

Details are given in the following of the earlier methods of strength design, as well as of the modern approach in which the buckling capacity of the conduit wall is explicitly investigated.

4.5.1 Empirical method of deflection control

Strength design of the conduit wall through deflection control was suggested by Spangler in 1941 (Ref. 38). He concluded from his experimental investigation on structures with relatively small circular conduits that the conduit wall fails with a reversal of curvature when the diameter of the conduit is reduced by about 20 percent by the overburden load. This kind of failure is usually referred to as *snap-through failure.* Spangler also observed that the shortening of the rise of the conduit matched the elongation of the span at failure. Incorporating a "factor of safety" of 4, the maximum allowable value of the increase of the span, denoted as Δx, was recommended to be $0.05D$, where D is the diameter of the circular conduit.

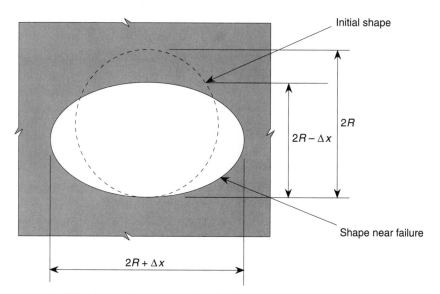

Figure 4.28 Pipe deformations assumed in the Iowa formula.

The following equation, known as the Iowa formula, was proposed for calculating Δx, which is considered to be the same as the shortening of the rise, as shown in Fig. 4.28.

$$\Delta x = \frac{D_1 D_2 W_c R^3}{EI + 0.061E' R^3} \tag{4.35}$$

where Δx is in inches; D_1 = the bedding constant which depends upon the bedding angle; D_2 = deflection lag factor; W_c = vertical load per unit length of the conduit, obtained by the equation following; R = mean radius of the conduit in inches; E = modulus of elasticity of conduit wall material in lb/in^2; I = moment of inertia of the conduit wall per unit length in in^4/in; and E' = modulus of soil stiffness in lb/in^2. It is noted that the original paper for this method, i.e., Ref. 38, refers to E' as the modulus of soil reaction.

The vertical load per unit length of the conduit, W_c, is given by

$$W_c = P_0 2R \tag{4.36}$$

where P_0 is the free-field vertical pressure at the crown in lb/in^2.

The passive pressure p activated at the sides of the conduit is given by

$$p = E' \frac{w}{R} \tag{4.37}$$

where w = radial displacement at the interface of the pipe and the soil, and consequently w/R is the radial strain in the soil. Denoting E'/R by k_n, the coefficient of soil reaction, Eq. (4.37) can be written as

$$p = k_n w \tag{4.38}$$

With this change of notation, Eq. (4.35) is often written in the following alternative form, in which $2D_1 D_2$ is replaced by 0.167 for normal conditions (Ref. 24).

$$\Delta x = \frac{0.167 P_c R^4}{EI + 0.061 k_n R^4} \tag{4.39}$$

Chapter 7 provides guidance on the values of E' and k_n.

Reference 44 contains the details of an experimental investigation which shows that the conduit wall can fail according to the Spangler criterion only when the soil and conduit wall have certain stiffnesses. Under different conditions, the conduit wall can fail well before Δx reaches the value of $0.2 D_h$. It was also observed in this reference that the maximum deflection ratio ($\Delta x/R$) that could be attained before failure by buckling increases with the increase of the ring stiffness parameter, $EI/(\rho_0 R^4)$, where ρ_0 is the density of the soil in contact with the conduit wall.

This stiffness parameter has in the past proved to be useful for quantifying the failure of small pipes having diameter of less than 1 m, but to be progressively less suitable for larger pipes. In particular, for pipe diameters greater than 3 m, the structures should be designed with the help of more rigorous strength analysis techniques.

4.5.2   Stability analysis of pipe embedded in elastic continuum media

The elastic continuum model has been applied by Forrestal and Hermann (Ref. 14) and by Moore (Refs. 31 through 34) to investigate the buckling of long cylindrical shells surrounded by elastic media. The loading exerted by the elastic medium on the shells in the buckling configuration was found by solving boundary value problems of the linearized theory of elasticity in the presence of initial stress. Formulas for the buckling pressure of the shells were derived and found to be a function of $(EI)^{1/3}$ multiplied by $(E_S^*)^{2/3}$, in which EI is the bending rigidity of the conduit wall and E_S^* is the secant modulus of the soil:

$$A f_e = 1.3(EI)^{1/3}(E_S^*)^{2/3} \qquad (4.40)$$

Equation 4.40 accounts for the elastic medium to provide equal resistance against the formation of a buckling wave with displacements toward the soil or away from it. This also means that shear forces are transferred through the elastic medium. It is to be noted that Eq. (4.40) is not a function of the conduit's radius of curvature R, since it leads to the development of short buckling waves along the conduit wall.

References 32 to 35 extended the use of the continuum elastic medium to the analysis of conduits with elliptic shapes, and modified Eq. (4.40) by applying a shape factor, R_s. Other shapes, such as pipe-arches, are not yet examined using the continuum model and may be treated using a rational approximate approach based on the shape factors of elliptic conduits.

It has been observed by Forrestal and Hermann (Ref. 14) and by Moore (Ref. 31) that the continuum model predicts pressures which are often too high. Moore accounted for this fact by applying a factor, $\Psi = 0.55$, to Eq. (4.40) in order to bring the theoretical results within a reasonable range of reported test results (Ref. 16). Equation (4.40) is modified as follows:

$$A f_e = 1.3\Psi(EI)^{1/3}(E_S^*)^{2/3}R_S R_h \qquad (4.41)$$

in which R_h is a factor to account for the effect of shallow cover.

The reported high theoretical results of Eqs. (4.40) and (4.41) can be attributed mainly to the fact that the preceding continuum analysis did not account for the actual manner in which loading is applied over the conduits. It is based on a linear formulation for the eigen value of uniform thrust over the conduit wall. However, the live load and the dead load due to soil cover constitute the active load which induces downward displacement of the upper zone of the conduit.

In order to simulate the actual load conditions, Ref. 1 conducted a stability analysis using a finite element program and an elastic continuum model in which the conduit is subjected to gravity dead load of soil cover, as well as to axle live load at the embankment. A comparison with the results of Ref. 1 show that Eq. (4.41) overestimates the buckling load because it does not account for the actual way the load is applied.

4.5.3   Stability analysis of elastically supported pipes

Booy (Ref. 9) appears to be the first investigator to have examined the possibility of failure of the conduit wall due to instability in compression, which leads to the development of buckling waves. To investigate such buckling behavior, he idealized the conduit wall as an axially loaded beam which is laterally restrained by elastic supports that are capable of transmitting both tension and compression. Using this approach, it was found that the elastic buckling, or critical, stress f_e in the conduit wall was given by

$$f_e = \frac{2\sqrt{EIk_n}}{A} \tag{4.42}$$

where A is the cross-sectional area of the conduit wall per unit length and the other symbols are as defined earlier.

Meyerhof and Baikie (Ref. 29) later examined the buckling of curved plates bearing against the soil. They observed that the number of buckling waves increased with increase in the value of the coefficient of soil reaction, k_n. When the number of these waves becomes large enough, the buckling stress, which then becomes independent of the length of the plate, is given by

$$f_e = \frac{2}{A} \left[\frac{EI\, k_n}{1 - \mu_c^2} \right]^{0.5} \tag{4.43}$$

where μ_c is the Poisson's ratio of the conduit wall material. The term $(1 - \mu_c^2)$ in the denominator of the right-hand side of Eq. (4.43) accounts for the effect of lateral strains in solid plates; as shown in Ref. 3, its value is closer to unity for corrugated plates, in which case Eq. (4.43) becomes the same as Eq. (4.42). It is noted that Eq. (4.43), like Eq. (4.42), is independent of the curvature of the conduit wall and is valid for relatively flexible conduit walls supported by a backfill of relatively high coefficient of subgrade reaction. The bending stiffness of the conduit wall relative to the stiffness of the backfill is characterized by a parameter α which is defined as

$$\alpha = \left[\frac{k_n R^4}{EI} \right]^{0.25} \tag{4.44}$$

For Eqs. (4.42) and (4.43) to be valid it is required by Ref. 29 that $\alpha \geq 2.0$; also, it is recommended that the following limits be adopted for arches.

$\alpha \geq \dfrac{\pi}{\delta}$ \qquad for arches with feet fixed against notation

$\alpha \geq \dfrac{1.5\pi}{\delta}$ \qquad for arches with hinged feet

where δ is one-half the center angle of the arch, measured in radians, as illustrated in Fig. 4.29.

Equation (4.43) was derived for circular pipes with the assumption of uniform soil support. This assumption is valid only when the depth of cover has a certain minimum value, which is assumed to be the pipe diameter. For structures in which the depth of cover is less than the pipe diameter, the buckling stress can still be determined by Eq. (4.43), but by using a reduced value of the modulus of soil reaction; the revised expression for f_e then becomes

$$f_e = \frac{2}{A} (EI \, \bar{k}_n)^{0.5} \tag{4.45}$$

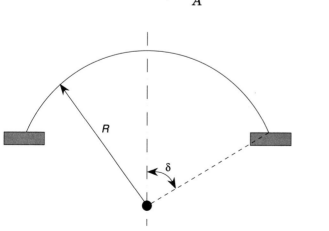

Figure 4.29 Definition of δ.

where $\bar{k}_n$, the modified modulus of soil reaction, as will be shown in Chap. 7, is given by

$$\bar{k}_n = \left\{ 1 - \left(\frac{R}{R+H} \right)^2 \right\} k_n \tag{4.46}$$

As recommended in Ref. 29, the buckling stress for the conduit wall under shallow fill is further reduced to account for the nonuniform distribution of the soil support. For structures with shallow covers,

$$f_e = \frac{2\psi}{A} (EI\ \bar{k}_n)^{0.5} \tag{4.47}$$

where $\psi = 1$ for $H/R \geq 1.0$, and is given by the following expression for $0.1 < H/R < 1.0$.

$$\psi = \left(\frac{H}{R} \right)^{0.5} \tag{4.48}$$

References 26 and 27 give details of the elastic analysis of buckling of long tubes with radial elastic supports subjected to uniform hydrostatic loading. It is shown in these references that the critical applied pressure p^* causing failure is given by

$$p^* = 2 \left[\frac{E'EI}{R^3} \right]^{0.5} \tag{4.49}$$

Since, as shown earlier

$$f_e = \frac{p^* R}{A} \tag{4.50}$$

and

$$k_n = \frac{E'}{R} \tag{4.51}$$

it is readily verified that Eq. (4.49) is the same as Eq. (4.42).

4.5.4 Analogy between the buckling of columns and conduit walls

For a soil-pipe system in which the soil is assumed to be noncompressible and with an angle of friction θ tending to zero, the radial pressure on the pipe is very nearly hydrostatic. In this case, as shown in standard text books (e.g., Ref. 42), f_e is given by

$$f_e = \frac{3E}{(R/r)^2} \tag{4.52}$$

where r is the radius of gyration of the conduit wall. Eq. (4.52), because of being based on the assumption of $\theta = 0$, underestimates the buckling stress; this is discussed in Ref. 45. The difference between the actual and assumed fluid soil can be accounted for by a soil stiffness parameter, K, in the same way as the effective length of a column is catered for in the equation for column buckling. Consequently, Eq. (4.52) can be modified as

$$f_e = \frac{3E}{(KR/r)^2} \tag{4.53}$$

It is suggested in Ref. 44 that the value of K is governed by the angle of friction θ of the soil. The stiffness factor K is assumed to be the ratio of horizontal to vertical soil pressure due to vertical loading on laterally restrained soil. Thus, for cohesionless soil,

$$K = \frac{1 - \sin \theta}{1 + \sin \theta} \tag{4.54}$$

The elastic buckling stress f_e given by Eq. (4.53) is plotted in Fig. 4.30 against the nondimensional factor (KR/r) which is referred to as the *ring flexibility factor*. It can be appreciated from this figure that the buckling stress given by Eq. (4.53) is analogous to the stress corresponding to the classical Euler buckling load for columns. It can also be appreciated that the buckling stress given by Eq. (4.53) can be valid only for large values of (KR/r). For smaller values of the ring stiffness factor, the yield stress of the conduit wall material may be smaller than f_e. For such cases, a transition has to be assumed between the yield stress f_y and f_e. Such transition curves, which are also used for column buckling equations, are shown in Fig. 4.30 for three different values of f_y; they are derived by assuming that the transition from elastic buckling to inelastic buckling takes place at $0.5f_y$. The stress of $0.5f_y$, according to Eq. (4.53), corresponds to the ring flexibility factor being equal to $\sqrt{6E/f_y}$. For ring flexibility below $\sqrt{6E/f_y}$, the failure stress is expressed by the parabolic transition curve, and is given by

$$f_b = f_y - \frac{f_y^2}{12E} \left(\frac{KR}{r} \right)^2 \tag{4.55}$$

The soil stiffness parameter, K, proposed in Ref. 44, i.e., that defined by Eq. (4.54), depends only upon the properties of soil. However, it can be appreciated that this factor should also depend upon the relative stiffness of the soil with respect to the rigidity of the conduit wall, which is characterized by EI/R^3, with EI being the flexural rigidity of

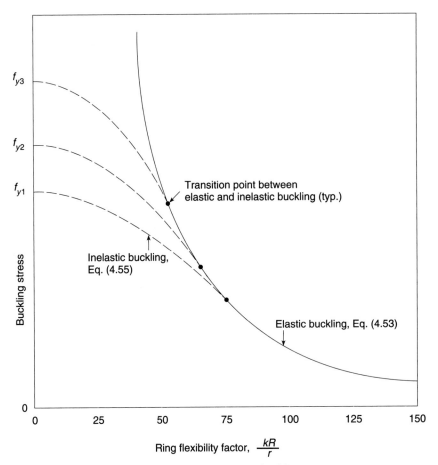

Figure 4.30 Transition between elastic and inelastic buckling stresses.

the conduit wall and R its radius of curvature. By equating the right-hand sides of Eq. (4.53) and (4.42), it can be shown that

$$K = B\left(\frac{EI}{E'R^3}\right)^{0.25} \tag{4.56}$$

in which B takes the value 1.5. It is recalled that $k_n = E'/R$ and that the term $(1 - \mu_c^2)$ is dropped because it is very close to unity.

The AASHTO specifications (Ref. 6) and AISI handbook (Ref. 7) provide expressions for f_b which are based upon Eq. (4.53), with the values of K being taken as 0.22 and 0.27, respectively. These expressions also account for the transition between elastic and inelastic buckling. However, the transition is assumed between f_u and $0.5f_u$ with f_u being the ultimate stress of the conduit wall material. This transition is illustrated conceptually in Fig. 4.31. Both the AASHTO and AISI expressions are valid only if the backfill is compacted to a minimum of 85 percent standard Proctor density. The AISI expressions for f_b, which are based upon $f_y = 33000$ psi and $f_u = 40000$ psi, are as follows in U.S. customary units with f_b being in psi.

$$
\left.
\begin{aligned}
f_b &= 33000 & \text{for } \frac{D_h}{r} &< 294 \\[2ex]
f_b &= 40{,}000 - 0.081\left(\frac{D_h}{r}\right)^2 & \text{for } 500 \geq \frac{D_h}{r} &\geq 294 \\[2ex]
f_b &= \frac{4.93 \times 10^9}{(D_h/r)^2} & \text{for } \frac{D_h}{r} &> 500
\end{aligned}
\right\} \quad (4.57)
$$

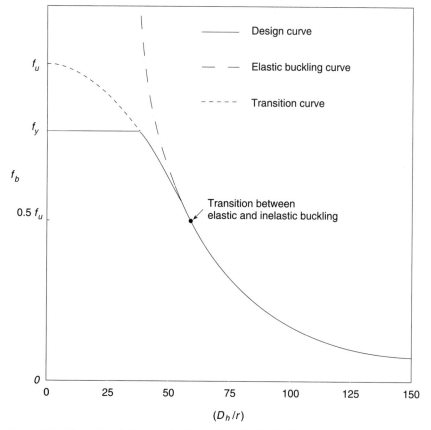

Figure 4.31 Transition between elastic and inelastic buckling stresses according to AASHTO specifications and AISI handbook.

The AASHTO specifications require that f_b should be the smaller of f_y and the value given by the relevant of the following expressions.

$$\left.\begin{array}{ll} f_b = f_u - \dfrac{f_u^2}{48E}\left(\dfrac{K D_h}{r}\right)^2 & \text{for } D_h < \dfrac{r}{K}\sqrt{\dfrac{24E}{f_u}} \\[4ex] f_b = \dfrac{12E}{(K D_h/r)^2} & \text{for } D_h > \dfrac{r}{K}\sqrt{\dfrac{24E}{f_u}} \end{array}\right\} \qquad (4.58)$$

4.5.5 Effect of conduit wall displacement on buckling stress

The Eqs. (4.42), (4.43), and (4.49) are similar to each other since they are all based upon the analysis of pinned circular arches in which the pin supports are fully restrained against tangential movement, as is illustrated in Fig. 4.32. It is noted that this boundary condition was also reproduced faithfully in the various test models that were used to verify the preceding equations. As discussed following, the assumption of unyielding supports is not realistic for the top portion of the pipe.

The effect of nonyielding supports on the buckling stress of the circular arch is discussed in Ref. 21, in which the pipe is divided into two zones. In one zone, the conduit wall moves toward the inside of the conduit under applied loads; this zone, illustrated in Fig. 4.33 as zone 1, corresponds to the conduit wall in the top portion of the pipe. In the other zone, the conduit wall moves toward the soil under applied loading; as also illustrated in Fig. 4.33, the side and bottom portions of the pipe fall into this second zone. The demarcation between the two zones is defined by one-half of the subtended angle θ_0, which is also illustrated in Fig. 4.33.

It has been shown in Ref. 21 that the assumption of yielding supports in the buckling analysis is valid only when application of the load results in the conduit wall moving toward the soil. By contrast, when application of the load causes the conduit wall to move away from the soil, as it does for the upper portions of the pipe, the assumption of

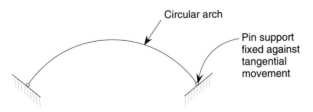

Figure 4.32 Illustration of boundary conditions used in some buckling analyses.

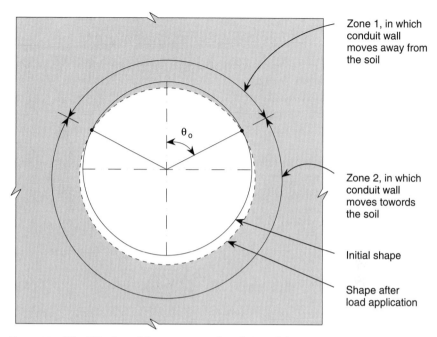

Figure 4.33 Identification of the two zones of conduit wall for buckling analysis.

unyielding supports is not realistic. In the latter case, the buckling analysis is carried out by assuming that the pin ends of the arch are supported by tangential springs, as is shown in Fig. 4.34d. The effect of tangential movement of the pin ends in the upper portion of the pipe is to reduce the buckling stresses.

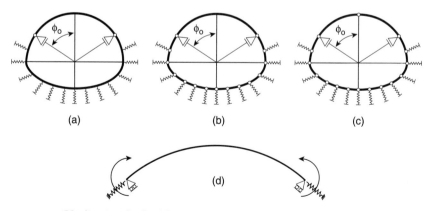

Figure 4.34 Idealization for buckling analysis of the top portion of the pipe: (a) two-hinged upper arch and rigid lower zone (upper buckling limit); (b) two-hinged upper arch and hinged lower zone; (c) three-hinged upper arch and hinged lower zone (lower buckling limit); (d) elastic support condition for the upper arch.

The transition between the two buckling zones, defined by θ_0, depends upon the flexibility factor $EI/k_n R^4$ and can be determined by idealizing the pipe as a plane frame supported by elastic springs. As discussed earlier in this section, the plane frame is analyzed under the sinusoidal active pressure shown in Fig. 4.16, and the springs in tension are eliminated through an iterative process.

It is noted that in determining θ_0, the principal objective is to divide the pipe roughly into two zones in which the conduit wall may have different flexural rigidities. Consequently, it is not necessary to determine θ_0 with any great precision, especially when the buckling analysis is carried out by simplified approximate methods. The lack of sensitivity of the buckling stress f_b to θ_0 can be appreciated readily if it is understood that f_b depends upon the effective rigidity of the whole zone rather than on the rigidity of a small segment. The value of θ_0 can be determined from the following approximate expression, which is given in Ref. 16.

$$\theta_0 = 1.6 + 0.2 \log\left(\frac{EI}{k_n R^4}\right) \tag{4.59}$$

where θ_0 is in radians.

The top portion of the pipe can be idealized either as a two- or three-hinge arch, as shown in Fig. 4.34a,b,c, respectively. Also as shown in Fig. 4.34, the lower portion of the pipe may be idealized as a frame or as a sprocket chain supported by springs as shown in Figs. 4.34a and 4.35a,b, respectively. If the pipe has the same flexural rigidity and soil support in the two zones, then the analysis of the top portion, idealized as a two-hinge arch supported by tangential and rotational springs, gives the upper-bound solution of the load-carrying capacity of the conduit. By contrast, the analysis of the lower portion, idealized as a spring-supported sprocket chain, gives the lower bound. It is expected that the actual capacity of the whole pipe lies between these two bounds.

The stiffness of the tangential spring support of the arch, which represents the top portion of the pipe, is governed by the elastic deformations of the lower portions of the pipe; these, in turn, are also governed by the coefficient of soil reaction k_n. When k_n has a very high value, the spring stiffness also becomes very large; for this condition f_e can be obtained by assuming the stiffness of the tangential support to be infinity, i.e., Eq. (4.45) or (4.53) can be used to obtain f_e for the case under consideration. In real life, the stiffness of the tangential supports is such that considerable tangential movement takes place, thus reducing f_e considerably below that obtained by Eq. (4.45) or (4.53). It is noted that for zero-support stiffness the arch becomes a mechanism as a result of which f_e becomes zero (Ref. 4).

It is recalled that for a given pipe, the buckling stress f_e is directly related to the critical radial pressure by Eq. (4.50). A nondimensional-

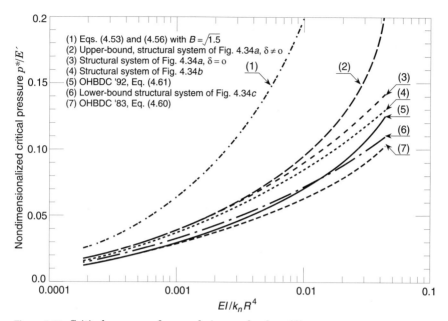

Figure 4.35 Critical pressures for round pipes under deep fill.

ized form of this critical pressure can be obtained by dividing it by the coefficient of soil reaction E'. The values of this nondimensionalized radial pressure p^*/E' in structures with deeply buried round pipes were calculated in Ref. 21 for different values of the flexibility factor $EI/k_n R^4$ by explicitly accounting for the stiffness of the tangential supports for the top arch. These values of p^*/E' are plotted in Fig. 4.35 along with the corresponding values obtained by using Eqs. (4.53) and (4.56) with B taken as $\sqrt{1.5}$. It can be seen in this figure that the difference between the two values of nondimensionalized critical pressure obtained respectively by these equations and by rigorous analysis widens for increasing values of the flexibility factor. This observation suggests that the stiffness factor K used in Eq. (4.53) and defined by Eq. (4.56) for $B = \sqrt{1.5}$, does not adequately reflect the relative flexibility of the conduit wall with respect to the soil envelope.

It was suggested in Ref. 4 that the stability formula be based on the lower limit, i.e., by considering the structural system shown in Fig. 4.34c, or curve 6 in Fig. 4.35. This can be expressed using Eq. (4.56) if the following expression is used for calculating the magnitude of B for the analysis of the upper portion of the pipe.

$$B = 1.22 \left\{ 1 + 2\left(\frac{EI}{E'R^3} \right)^{0.25} \right\}$$

(4.60)

The results of this formula are presented as curve 7 in Fig. 4.35. However, inelastic buckling analysis of frames on elastic supports (Ref. 15) led to the conclusion that, in general, plastic hinges start to develop at the crown and then at the upper haunches before failure develops. Plastic hinges usually do not develop in the lower zone of the conduit up to the failure load. That is to say, using the lower limit is too conservative and the magnitude of the parameter, B, may be modified (OHBDC, 1992) as follows:

$$B = 1.22\left\{1 + 1.6\left(\frac{EI}{E'R^3}\right)^{0.25}\right\}$$ (4.61)

which leads to the results expressed by curve 5 in Fig. 4.35. Reference 2 confirms that for the side and lower portions of the pipe, B should still be taken as $\sqrt{1.5} \simeq 1.22$.

4.5.6 Multispan conduits

Multispan conduits are often built with arbitrary required minimum spacing which differs considerably from one code to another. Narrow spacing between conduits causes the support, or stiffness, of the soil between the conduits to be different from the exterior sides of the conduit (Fig. 4.36). This phenomenon occurs because the soil between conduits transferring pressure to the unloaded conduit may be displaced to a degree larger than equivalent mass of soil.

An analysis using a plane finite element program has been conducted (Ref. 1) for the buckling of a conduit with live load limited to its domain (Fig. 4.36). The results obtained are illustrated in Fig. 4.37,

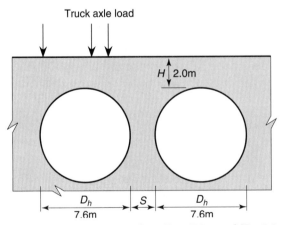

Figure 4.36 Twin conduits, where $D_h = 7.6$ m and $H = 2.0$ m.

which shows that the buckling load is reduced with the reduction of spacing down to 75 percent of the buckling load for a spacing ratio S/D_h = 0.1.

However, it should be noted that these results are obtained from a plane strain analysis and do not account for the dispersion of live load in the third dimension. Therefore, it is felt that such an analysis overestimates the reduction in the buckling load. The OHBDC (Ref. 30) applies a reduction factor:

$$F_m = (0.85 + 0.3S/D_h) \le 1.0 \tag{4.62}$$

F_m may be applied to reduce the magnitude of the buckling stress f_b calculated previously.

4.6 Strength Analysis of Soil Cover

As discussed in Sec. 4.2, the soil above the conduit can fail either by wedge sliding, as illustrated in Fig. 4.6a, or through tension, as illustrated in Fig. 4.6b; the analysis of these two modes of failure is discussed in this section.

4.6.1 Wedge sliding failure

Figure 4.38a shows the cross section of a soil-steel bridge carrying a strip loading which is of width b and is continuous in the direction perpendicular to the cross section. If the conduit wall of this bridge were strong enough to resist failure by buckling, the failure of the structure could take place by sliding of a wedge of soil situated on the right-hand side of the load; this wedge is identified as soil mass A in Fig. 4.38a,

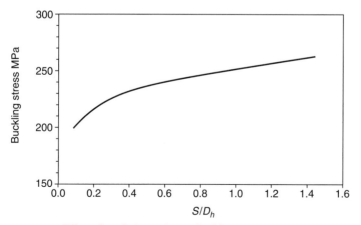

Figure 4.37 Effect of conduit spacing on buckling stress.

and the soil mass directly under loading is B. Figure 4.38b shows schematically the familiar potential failure surface corresponding to wedge-sliding failure of the foundation for a case in which a continuous strip footing rests on the ground. It can be seen in these two figures that there is a striking resemblance between the modes of failures of the two structures.

Simply by relying on engineering judgment it can be predicted that, because of the presence of the flexible conduit, the failure load for the bridge shown in Fig. 4.38a should be smaller than the failure load for the footing shown in Fig. 4.38b. The effect of the presence of the conduit wall on the wedge-sliding type of failure has been studied analytically in Ref. 21. This study involved consideration of the equilibrium of the two masses of soil identified as A and B in Fig. 4.38a. The graphical design aids given in Ref. 21 are limited to round pipes and pipe-arches having spans up to 4.0 m. However, they are still useful in

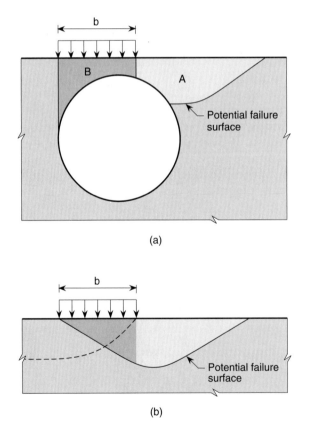

(a)

(b)

Figure 4.38 Wedge sliding failures: (a) soil-steel bridge; (b) shallow foundation.

developing a "feel" for the failure load for wedge sliding, as may be seen in the following example.

The example considered is that of a structure with an infinitely long, round conduit of 2.0 m diameter and a depth of cover of 0.25 m. The soil around the conduit is granular, with a 30° angle of internal friction. The corrugation profile for the conduit wall is 152 × 51 mm (6 × 2 in). As shown in Fig. 4.39, the structure is subjected to a uniformly distributed load whose width b, measured along the conduit span, is 1.26 m, and which is infinitely long along the conduit axis. From the information presented in Ref. 21, it can be shown that the pressure p_u causing wedge-sliding failure is given by

$$p_u = F \times \gamma \times b \tag{4.63}$$

where γ is the unit weight of the soil, b is the width of the loading strip, and F is a dimensionless coefficient whose value depends upon the thickness of the conduit for a given structure and corrugation profile. Reference 21 provides the information needed for calculating the values of F for conduit wall thickness of 2.5, 4.0, 5.5, and 7.0 mm, respectively, for the structure under consideration. These values of F are plotted against conduit wall thickness in Fig. 4.40. The points representing the discrete values of F are joined by a continuous curve which illustrates quite effectively the influence of the conduit wall thickness, and consequently of its flexural rigidity, on the load for wedge-sliding failure. It can be seen in this figure that, as expected, the failure pressure approaches zero as the conduit wall thickness approaches zero. From extrapolation it can be concluded that any increase in the conduit wall thickness beyond 9 mm has a negligible effect on the failure pressure.

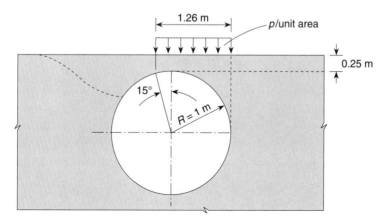

Figure 4.39 Strip loading on a structure with round conduit.

The familiar Terzaghi expression for ultimate bearing capacity of p_u of a long, continuous foundation placed above ground is as follows:

$$p_u = 0.5b\ \gamma N\gamma + c\ N_c \qquad (4.64)$$

where $N\gamma$ and N_c are coefficients depending upon the angle of internal friction and on the soil cohesion, respectively. For the case under consideration, the cohesion, c, is assumed to be zero and the value of $N\gamma$ is found to be 19.7. This gives $p_u = 9.85b\ \gamma$, i.e., $F = 9.85$. It is interesting to note that, as shown in Fig. 4.40, this value of F is fairly close to the upper-bound value of F corresponding to a foundation with a conduit running through it. The fact that the former value of F is somewhat smaller than the latter upper-bound value may be attributed either to the approximate nature of the analyses which yield these results or to the possibility that a very stiff embedded pipe can enhance the wedge-sliding failure load of a foundation.

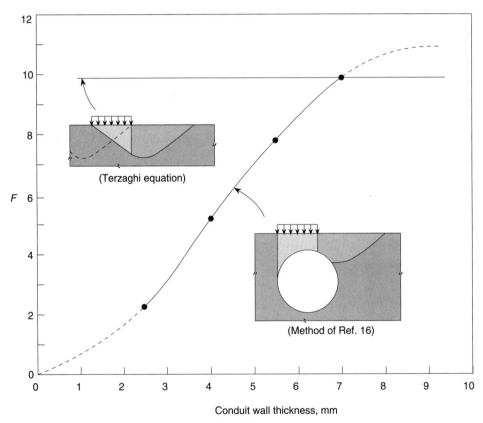

Figure 4.40 Coefficient F defining sliding wedge failure.

It has been observed in Ref. 21 that structures with round pipes and pipe-arches of spans less than about 3.5 m (11.5 ft) are not likely to experience wedge-sliding failure of the soil cover even when the cover is relatively shallow and the angle of friction of the soil is small. Wedge-sliding failure need not also be considered for structures with larger conduits provided that they comply with the minimum depth of cover requirements which will be discussed in Chap. 5.

It is only for exceptional structures that an analysis is required for wedge-sliding failure, it being noted that a structure needing such analysis would have: (1) an unstiffened conduit wall in the top portion with an unusually large radius of curvature; and (2) poor-quality soil cover with a very shallow depth. For such exceptional structures, the method of Ref. 21 can be used. However, it should be noted that this method applies only to continuous strip loadings, for which the failure load is usually smaller than that for loadings which have a finite length measured along the conduit span.

4.6.2 Tension failure

Granular, noncohesive soils are considered to have no tensile strength. The development of tension in the soil cover can trigger the failure of the entire soil-steel bridge. It can be appreciated intuitively that tension is more likely to develop in a shallow soil cover than in a deep one. Usually, the occurrence of tension failure in the soil is prevented by providing a sufficient depth. The depth of cover requirements of the AASHTO specifications (Ref. 6) and the AISI handbook are empirically based. However, the Ontario code requirements were derived from the analytical studies reported in Ref. 18. These studies used finite element analyses to establish a criterion for the failure of soil cover by tension or shear. The finite element analyses were based upon plane-strain idealizations incorporating nonlinear, stress-dependent, soil properties according to Ref. 47. Failure in the soil elements was established by using the Mohr-Coulomb's criterion.

In these studies, the failure of the soil cover was examined for a single-axle load placed on the embankment above the conduit. The load causing failure was denoted as p_f. Details of these analyses are given in Refs. 17 and 18, and the major conclusions drawn from these studies are presented in the following. For the sake of convenience, the failure load is nondimensionalized by dividing it by a load $p_a = 142$ kN (32 kips). It is recalled that this is one of the axle loads of the AASHTO vehicle, which is shown in Fig. 4.4a.

Fig. 4.41 shows the variation of the nondimensionalized failure load plotted against H/D_h for three specific conduit shapes having span-to-rise ratios of 0.63, 1.00, and 1.67, respectively. It can be seen from the results plotted in this figure that (1) the failure increases rapidly with increase in the depth of cover $H;$ and (2) the failure load decreases with increase in the span-to-rise ratio of the conduit. These observations are also confirmed by the results plotted in Fig. 4.42, which illustrates the effect of the conduit span on the failure load for a given depth of cover.

The preceding observations, which are entirely as expected, cast serious doubts on the practice of specifying a minimum depth of cover without paying any regard to the shape of the conduit. From the work presented in Refs. 1 and 18, there emerged naturally the requirement that the minimum depth of cover in a soil-steel bridge should be the larger of $\dfrac{D_h}{6}\left(\dfrac{D_h}{D_v}\right)^{0.5}$ and $0.4\left(\dfrac{D_h}{D_v}\right)^2$ meters, where D_h and D_v are in meters. As will be discussed in Chap. 5, this requirement has been adopted by the Ontario Code (Ref. 30).

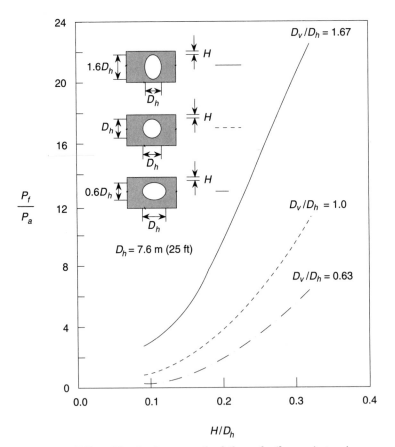

Figure 4.41 Effect of depth of cover on the failure of soil cover in tension.

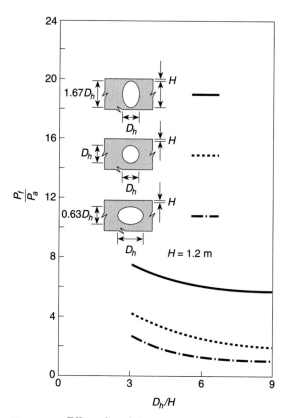

Figure 4.42 Effect of conduit size on the failure of soil cover in tension.

References

1. Abdel-Sayed, G., and Girges, Y., "Stability of Soil-Steel Bridges," *Canadian Journal of Civil Engineering,* vol. 19, no. 3, June 1992, pp. 463–468.
2. Abdel-Sayed, G., Bakht, B., and Selig, E., "Soil-Steel Structures Design by the 3d Edition of OHBDC, "*Canadian Journal of Civil Engineering,* vol. 19, no. 4, August 1992, pp. 545–550.
3. Abdel-Sayed, G., "Critical Shear Loading of Curved Panels or Corrugated Sheets," *Journal of the Engineering Mechanics Division,* ASCE, 96 (EM6), 1970, pp. 895–912.
4. Abdel-Sayed, G., "Stability of Flexible Conduits Embedded in Soil," *Canadian Journal of Civil Engineering,* 5(33), 1978, pp. 324–334.
5. Abdel-Sayed, G. and Bakht, B., "Analysis of Live-Load Effects in Soil-Steel Structures," *Transport Research Board Records,* no. 878, 1983, pp. 49–55.
6. American Association of State Highway and Transportation Officials (AASHTO), *Standard Specification for Highway Bridges,* Washington, D.C., 1989.
7. American Iron and Steel Institute (AISI), *Handbook of Steel Drainage and Highway Construction Products,* Washington, D.C., 1983.
8. Bakht, B., "Soil-Steel Structure Response to Live Loads," *ASCE Journal of the Geotechnical Division,* 1981, pp. 779–798.

9. Booy, C., "Flexible Conduit Studies," Prairie Farm Rehabilitation Administration, Canada Department of Agriculture, Saskatoon, Saskatchewan, 1957.
10. Burns, J. Q., "An Analysis of Circular Cylindrical Wells Embedded in Elastic Media," Ph.D. thesis, University of Arizona, Tucson, AS., 1965.
11. Canadian Standards Association (CSA), CAN/CSA-56-88, *Design of Highway Bridges,* Rexdale, Ontario, Canada, 1988.
12. Duncan, J. M., "Behaviour and Design of Long-Span Metal Culvert Structures," *ASCE Journal of the Geotechnical Division,* 105(GT3), 1979, pp. 399–417.
13. Duncan, J. M., and Chang, C. Y., "Nonlinear Analysis of Stress and Strain in Soils," *ASCE Journal of Mechanics and Foundation Division,* 96(SM5), 1970, pp. 1629–1653.
14. Forrestal, M. J., and Herrmann, G., "Buckling of a Long Cylindrical Shell Surrounded by an Elastic Medium," *Int. J. Solids Struct.,* vol. 1, 1965, pp. 297–310.
15. Ghobrial, M., and Abdel-Sayed G., "Inelastic Buckling of Soil-Steel Structures," *Transportation Research Record 1008,* Washington, D.C., 1985, pp. 7–14.
16. Gumbel, J. E., "Analysis and Design of Buried Flexible Pipes," Ph.D. thesis, Department of Civil Engineering, University of Surrey, U.K., 1983.
17. Hafez, H. H., "Soil-Steel Structures under Shallow Cover," Ph.D. thesis, Department of Civil Engineering, University of Windsor, Windsor, Ontario, 1981.
18. Hafez, H., and Abdel-Sayed, G., "Soil Failure in Shallow Covers above Flexible Conduits," *Canadian Journal of Civil Engineering,* 10(4), 1983, pp. 654–661.
19. Haggag, A. A., "Structural Backfill Design for Corrugated-Metal Buried Structures," Ph.D. dissertation, Dept. of Civil Eng., Univ. of Massachusetts, 1989.
20. Katona, M. G., Smith, J. M., Odello, R. J., and Allgood, J. R., *CANDE: Engineering Manual*—"A Modern Approach for the Structural Design of Buried Culverts," FHWA/RD-77-5, Navy Civil Engineering Laboratory, Port Hueneme, Calif., 1976.
Katona, M. G., Smith, J. M., Odello, R. J., and Allgood, J. R., *CANDE: Engineering Manual*—A Modern Approach for the Structural Design of Buried Culverts," FHWA/RD-77-5, Navy Civil Engineering Laboratory, Port Hueneme, Calif., 1976.
21. Kloeppel, K., and Glock, D., "Theoretische und Experimentelle Untersuchungen zu den Traglastproblemen biegeweigher, in die Erde eingebetteten Rohre," publ. no. 10, Institut für Statik and Stahlbau, T. H. Darmstadt, Germany, 1970.
22. Krizek, R. J., Parmelee, R. A., Kay, J. N., and Elnaggar, H. A., "Structural Analysis and Design of Pipe Culverts," *Report No. 116,* National Cooperative Highway Research Program, Highway Research Board, Washington, D.C., 1971.
23. Lee, R.W.S., and Kennedy, D.J.L., "Behaviour of Bolted Joints of Corrugated Steel Plates, *Report for Alberta Transportation,* prepared by Dep. of Civil Engineering, University of Alberta, Alberta, Canada, 1988.
24. Leonard, G. A., Wu, T. H., and Juang, C. H., "Predicting Performance of Buried Conduits," *Report on Joint Highway Research Project No. C-36-62F,* Purdue University, 1982.
25. Leonards, G. A., and Juang, C. H., "Comparison between Predicted Response of Soil-Steel Structures with Design Requirements of the Ontario Highway Bridge Design Code," School of Engineering, Purdue University, 1983.
26. Link, H., "Bietrag zum Knickproblem des elastisch gebetteten Kreisbogentraegers," Stahlbau, 32(7), 1963.
27. Luscher, U., "Buckling of Soil-Surrounded Tubes," *J. Soil Mech. Found. Div.,* Proc. Am. Soc. Civ. Engrgs., vol. 92, no. SM6, Nov., 1966, pp. 211–228 (discussed in vol. 93 (1967): no. SM2, p. 163; no. SM3, pp. 179–183; no. SM5, pp. 337–340, author's closure in vol. 94 (1968), no. SM4, pp. 1037–1038).
28. Meyerhof, G. G., "Composite Design of Shallow-Buried Steel Structures," *Annual Convention of the Canadian Good Roads Association,* Halifax, N.S., 1966.
29. Meyerhof, G. G., and Baikie, L. D., "Strength of Steel Culvert Sheets Bearing Against Compacted Sand Backfill," *Highway Research Record no. 30,* Highway Research Board, National Academy of Science, Washington, D.C., 1963.
30. Ministry of Transportation of Ontario, "Ontario Highway Bridge Design Code (OHBDC), "Downsview, Ontario, Canada, 1983, 1992.
31. Moore, I. D., "The Stability of Buried Tubes," Ph.D. thesis, School of Civil and Mining Engineering, University of Sydney, Australia, 1985.

32. Moore, I. D., "The Elastic Stability of Shallow Buried Tubes," *Geotechnique,* vol. 27, no. 2, 1987, pp. 151–161.
33. Moore, I. D., "Elastic Buckling of Buried Flexible Tubes—A Review of Theory and Experiment," Dept. *Research Report 025.09.87,* of Civil Engineering and Surveying, University of Newcastle, NSW, Australia, 1987.
34. Moore, I. D., "Buckling of Buried Flexible Tubes of Noncircular Shape," *Int. Conf. on Numerical Methods in Geomechanics,* Innsbruck, April, 1988.
35. Moore, I. D., Selig, E. T., and Haggag, A., "Elastic Buckling Strength of Buried Flexible Culverts," *Pre-prints TRB Session 143,* Washington, D.C., 1988.
36. Poulos, H. G., and Davis, E. H., *Elastic Solutions for Soil and Rock Mechanics,* John Wiley & Sons, Inc., New York, N.Y. 1976.
37. Selig, E. T., "Subsurface Soil-Surface Interaction: A Synopsis," *Highway Research Record No. 413,* Highway Research Board, Washington, D.C., 1972.
38. Spangler, M. G., "The Structural Design of Flexible Pipe Culverts," *Bull. 153,* Iowa State College Engineering Experimental Station, Iowa, 1941.
39. Spangler, M. G., "Field Measurements of the Settlement Ratios of Various Highway Culverts," *Bull. 170,* Engineering Research Institute, Iowa State University, Ames, Iowa.
40. Spangler, M. G., *Soil Engineering,* 2nd ed., International Textbook Company, Scranton, Pa., chap. 24, 1964.
41. Szechy, K., *The Art of Tunnelling,* Publishing House of the Hungarian Academy of Science, 2nd English edition, Budapest, 1973.
42. Timoshenko, S. P., and Gere, J. M., *Theory of Elastic Stability,* McGraw-Hill, New York. 1961.
43. Valera, J. E., and Donovan, N. C., "Soil Liquifaction Procedures—a Review," *ASCE Journal of Geotechnical Engineering,* 103(GT6), 1977, pp. 607–625.
44. Watkins, R. K., "Failure Conditions of Flexible Culverts Embedded in Soil," Highway Research Board, *Proceedings of the Annual Meeting,* 39, 1960.
45. Watkins, R. K., "Structural Design of Buried Circular Conduits," *Highway Research Record No. 145,* pp. 1–16. 1966.
46. White, H. A., and Layer, J. P., "The Corrugated Metal Conduit as a Compression Ring," Highway Research Board, *Proceedings of the Annual Meeting,* 39, 1960, pp. 389–397.
47. Wong, S. K., and Duncan, J. M., "Hyperbolic Stress-Strain Parameters for Non-Linear Finite Element Analysis of Stresses and Movements in Earth Masses," *Geotechnical Engineering Report,* TE 74-3, University of California, Berkeley, California, 1974.

Chapter 5

Structural Design Procedures

George Abdel-Sayed
University of Windsor, Windsor, Ontario, Canada

Baidar Bakht
Ministry of Transportation of Ontario, Canada

Ron P. Parish
Bolter Parish Trimble Ltd., Alberta, Canada

It is very desirable that the designer of a soil-steel bridge become familiar with the philosophy of structural design discussed in Chap. 4. However, there may be some designers who would like to go directly to the details of the design procedures. For such designers, this chapter contains all the requirements of the following three specific design methods.

1. Ontario Code methods
2. AASHTO specifications method
3. Duncan method

These three methods have the following steps of calculations in common.

1. Determine the specified load effects in the conduit wall due to the design live and dead loads.
2. Select the corrugation profile and the thickness of the conduit wall so that its strength is adequate to sustain the load effects determined in step 1.
3. Check that the selected conduit wall has adequate stiffness for handling during construction.

4. Select a bolting arrangement for longitudinal seams of the pipe so that its strength is adequate to sustain the conduit wall thrust calculated in step 1.

The requirements of each design method are linked to certain specified conditions, such as the minimum depth of cover, the spacing between conduits, and the extent of engineered soil. The various design methods are, therefore, valid only when used in conjunction with the respective conditions.

It is recalled that the soil cover above the conduit is not designed explicitly since its design is implicit in the requirements for minimum depth of cover.

Besides the details of the various design procedures, this chapter also contains design calculations corresponding to each of these procedures for the examples of two specific soil-steel bridges. Extensive discussion is provided in these examples at the various stages of the calculations, so as to enable the designer to use the given procedures for structures other than those dealt with in the examples.

5.1   Ontario Code Method

The Ontario Highway Bridge Design Code (OHBDC) was first introduced in 1979, its second edition was issued in 1983, and the third in 1992 (Ref. 4). This code contains a separate section for the design of soil-steel bridges, which are there called "soil-steel structures." The OHBDC is based upon the limit states design philosophy; its design vehicle consists of five axles, as shown in Chap. 4, Fig. 4.4d.

Since the OHBDC, in its entirety, is only in metric units, U.S. customary equivalents are not given in this section.

5.1.1   Design criteria

For the design of soil-steel bridges, the OHBDC requires only the consideration of the ultimate limit state (ULS) under one loading combination of dead and live loads, with the live loading including the effects of impact. For the design of the conduit wall, this code also requires that consideration be given only to the thrust in the conduit wall, with bending moments in it being neglected—the criterion for the structural design of the conduit wall for ULS can be expressed as

$$\phi R_n \geq \alpha_D T_D + \alpha_L(1 + DLA)T_L \qquad (5.1)$$

where ϕ = the resistance factor, being 0.80 for the strength of the conduit wall and 0.70 for the strength of longitudinal seams

α_D = the load factor for dead loads, being 1.25

α_L = the load factor for live loads, being 1.40

DLA = the dynamic load allowance which is specified to be 0.4 for zero depth of cover, decreasing linearly to 0.1 for a depth cover of 2.0 m; for depths of covers larger than 2.0 m, DLA is specified to be 0.1.

T_D = the conduit wall thrust due to unfactored dead loads

T_L = the conduit wall thrust due to unfactored design live loading which is shown in Chap. 4, Fig. 4.4d.

R_n = the nominal capacity of the conduit wall to sustain thrusts; this is given by

$$\phi R_n = A\, f_b \tag{5.2}$$

in which A = cross-sectional area of the conduit wall per unit length, it being noted that the values of A for 152 × 51 mm corrugated plates are given in Chap. 1, Table 1.3 for the plate sections generally available in Canada

f_b = the compressive failure stress in the conduit wall, which is calculated by the simplified method given later in this section

5.1.2 Dead-load thrust

The thrust T_D in the conduit wall is assumed to have the same value all round the conduit; it is obtained from

$$T_D = 0.5(1.0 - 0.1C_s)A_f W \tag{5.3}$$

in which W is the weight of the fill directly above the conduit per unit length as shown in Chap. 4, Fig. 4.1; A_f is a coefficient obtained from Fig. 4.22 depending upon the depth of cover and the shape of the conduit determined by dimensions D_h and D_v which are defined in Fig. 4.23. C_s is a dimensionless parameter defining the relative axial rigidity of the conduit wall with respect to the soil stiffness as follows.

$$C_s = \frac{E_s^*}{E}\,\frac{D_v}{A} \tag{5.4}$$

where E = modulus of elasticity of conduit wall material

E_s^* = effective secant modulus of soil which may be taken as $\overline{E}_s/(1 - \nu^2)$, or from Table 5.1 in the absence of test data

$\overline{E}_s$ = Young's modulus of soil

D_v = dimension relating to the cross-section of the conduit as defined in Chap. 4, Fig. 4.23

ν = Poisson's ratio of soil

TABLE 5.1   Values of E' and E_s^* for Various Soils

Soil group number	Standard Proctor density	Modulus of soil stiffness, E' MPa	Effective secant modulus of soil, E_s^*, MPa
I	Between 85 and 90%	7.0	13.0
	Between 90 and 95%	13.0	25.0
	Greater than 95%	22.0	40.0
II	Between 85 and 90%	3.0	5.0
	Between 90 and 95%	7.0	12.0
	Greater than 95%	13.0	20.0

5.1.3   Live-load thrust

The live-load thrust T_L is also assumed to have the same value all round the conduit and is obtained from

$$T_L = 0.5(\text{lesser } D_h \text{ and } l_t)\sigma_L m_f(1 + DLA) \qquad (5.5)$$

where σ_L, the equivalent uniformly distributed load at the crown level, is obtained by: (1) positioning the 320 kN tandem of the OHBD truck shown in Fig. 4.4d, centrally above the crown at embankment level, and (2) distributing the wheel loads through the fill in the span direction of the conduit at a slope of one vertically to one horizontally and in the longitudinal direction of the conduit at a slope of two vertically to one horizontally. When areas of dispersed loads overlap, the load is assumed to be distributed uniformly over the rectangular area subtended by the extremities of the individual dispersed areas. Such an overall rectangular area is shown in Fig. 5.1 for two specific cases. This figure also defines the length l_t of the dispersed load at crown level. The factor m_f is the modification factor to account for multilane loading; its value is specified to be 1.0, 0.9, 0.8, and 0.7 for one, two, three and four loaded lanes contributing to σ_L, respectively. D_h is the conduit span.

The term $\sigma_L m_f(1 + DLA)$ can be regarded as the effective static live load pressure at the crown level. Since the value of this pressure varies only with the depth of cover H, it can be plotted against H, as shown in Fig. 5.2. To facilitate reading, a table is also given along with this figure which gives the values of the effective pressure for specific values of H. It can be seen from Fig. 5.2 that for depths of cover larger than, say, 10 m, the effective live-load pressure at the crown level becomes negligible.

5.1.4   Compressive strength

The OHBDC requires that the strength of the conduit wall to sustain thrust be calculated from the simplified method which is reproduced

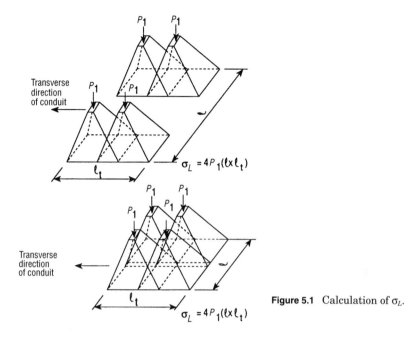

Figure 5.1 Calculation of σ_L.

later in this subsection for determining the failure compressive stress f_b. It is noted that f_b corresponds to both the elastic buckling stress and the crushing stress, with an assumed transition between the two which was discussed in Chap. 4, Sec. 4.5. For the method under consideration, the pipe is divided into upper and lower zones. As shown in Fig. 4.33, the "upper zones" consists of the pipe between the shoulders, and the "lower zones" includes the side segments and, if present, bottom segments. It is emphasized that it is not necessary to define the exact boundary between the top and bottom zones for the purpose of strength analysis.

The factored compressive failure stress f_b is calculated from the relevant of the following two equations

$$f_b = \frac{3\phi\rho F_m E}{(KR/r)^2} \qquad\qquad \text{for } R > R_e \qquad (5.6)$$

$$f_b = \phi F_m \left[f_y - \left\{ \frac{f_y^2}{12E\rho} \left(\frac{KR}{r} \right)^2 \right\} \right] \qquad R \le R_e \qquad (5.7)$$

where the equivalent radius of curvature R_e is defined as

$$R_e = \frac{r}{K} \left\{ \frac{6E\rho}{f_y} \right\}^{0.5} \qquad\qquad (5.8)$$

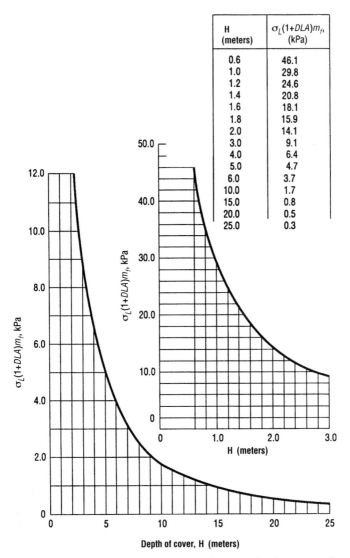

Figure 5.2 Effective live load pressure at the crown level corresponding to OHBDC.

ρ is a reduction factor accounting for the depth of cover H; it is given by

$$\rho = \left(\frac{H}{R_c}\right)^{0.5} , \quad \text{but} \not> 1.0 \tag{5.9}$$

As indicated, ρ is set $= 1.0$ when found > 1.0, i.e., when $H > R_c$.

The factor K, which depends upon the relative stiffness of the conduit wall with respect to the adjacent soil, is given by

$$K = \lambda \left\{ \frac{E I}{E_m} R^3 \right\}^{0.25} \tag{5.10}$$

where E_m is the modified modulus of soil stiffness which, like ρ, accounts for the depth of cover, and which is given by

$$E_m = E'\left\{ 1 - \left(\frac{R_c}{R_c + H}\right)^2 \right\} \tag{5.11}$$

λ is a factor which affects the value of K; for the lower portions of the pipe, its value is 1.22, and for the upper portions, it is given by

$$\lambda = 1.22 \left\{ 1.0 + 1.6 \left(\frac{E I}{E_m R_c^3}\right)^{0.25} \right\} \tag{5.12}$$

λ is also equal to 1.22 for the conduit walls of circular arches having rise-to-span ratios of less than 0.4.

For ready reference, all of the notation associated with buckling analysis which is not defined immediately preceding is defined as follows:

E = modulus of elasticity of conduit wall material
E' = modulus of soil stiffness
E_m = modified modulus of soil stiffness to account for the effect of shallow cover
F_m = 1.0 for single conduits, and $0.85 + 0.3S/D_h$ for multiple conduits, where S is the smallest clear spacing between adjacent conduits
f_y = yield stress of the conduit wall material
I = second moment of area of the cross section of the conduit wall about its longitudinal axis, per unit length
A = area of the cross section of the conduit wall, per unit length
R = radius of curvature of the conduit wall taken inside the conduit at the reference point on the cross section
R_c = radius of curvature of the conduit at the crown
r = radius of gyration of corrugation profile, $(I/A)^{0.5}$

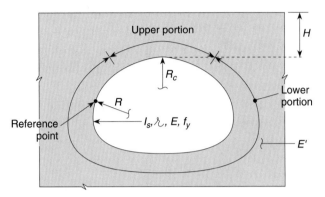

Figure 5.3 Notation used for buckling analysis at the conduit wall by OHBDC.

Some of the notation associated with the buckling analysis is also illustrated in Fig. 5.3.

In the absence of actual test data, Table 5.1 is permitted to be used for determining the values of E' for a given type of soil compacted to a known density. The soil group numbers referred to in this table are defined as follows:

Group I soils. These are coarse-grained soils exhibiting time-independent behavior; the moduli values are affected mainly by the degree of compaction. Some of the soils included in this group are listed in Table 5.2. It is noted that "granular A" and "granular B" are designations for two specific gradings of soils used in Ontario; these soils will be described in Chap. 7. It is recommended that only group I soils be used in the engineered backfill zone, the extent of which is discussed later in the section.

TABLE 5.2   Soil Classification for E' and E_s

Soil group number	Grain size	Soil types included	Unified soil classification symbol
I	Coarse	Well-graded gravel or sandy gravel	GW
		Poorly graded gravel or sandy gravel	GP
		Well-graded sand or gravelly sand	SW
		Poorly graded sand or gravelly sand	SP
		Granular A soil	GW, SW
		Granular B soil	SW, SP
II	Medium	Clayey gravel or clayey sandy gravel	GC
		Clayey sand or clayey gravelly sand	SC
		Silty sand or silty gravelly sand	SM

Group II soils. These soils are more finely grained than the group I soils; they are basically noncohesive in character but include some cohesive soils such as clayey sand and clayey gravel, as shown in Table 5.2. It is preferred not to use the soils in this group for the engineered backfill.

Group III soils. These soils are fine-grained cohesive soils of low plasticity. The 1983 edition of the OHBDC permitted the use of the soils of this group; however, the very low value of E' specified for them made their use impractical. Accordingly, this group of soils has been deleted from the latest edition of the OHBDC (Ref. 4). It is noted that fine-grained cohesive soils are not permitted to be used in the engineered backfill zone, and that the unified soil classification symbols used in Table 5.2 will be discussed in Chap. 7.

To assist designers, the compressive failure stress, f_b, calculated by the foregoing procedures is plotted for structures having circular conduits in Figs. 5.4, 5.5, 5.6, 5.7, and 5.8 corresponding to plate thickness of 3.0, 4.0, 5.0, 6.0, and 7.0 mm, respectively. It can be seen in these figures that f_b is plotted against the radius of curvature R for different values of H, and that its values are given separately for upper and lower portions of the pipe. These values of f_b are for $f_y = 230$ MPa and $E' = 13.0$ MPa, which is the value for well-graded gravel or sandy soils compacted to between 90 and 95 percent standard Proctor densities.

5.1.5 Seam strength

The bolted longitudinal seams of the pipe are only required to be designed to sustain the axial thrust. The nominal ultimate strength S_s of the longitudinal seams for the two bolting arrangements permitted by the OHBDC can be read directly from Fig. 5.9, corresponding to the thickness of the thinner of the two mating plates. Alternatively, the nominal strength can be read from Table 4.1 in Chap. 4. It is noted that these nominal strengths correspond to seams bolted with 20-mm-diameter bolts.

The strength of longitudinal seams should satisfy the following criterion:

$$\phi S_s \geq \alpha_D T_D + \alpha_L (1 + DLA) T_L \tag{5.13}$$

where the resistance factor ϕ for seam strength is 0.70.

It can be seen from Fig. 5.9 that, except for 7.0-mm plates, the designer does not have any choice in selecting a bolting arrangement. For plates thinner than 7.0 mm, the seam strength can be regulated only by changing the thicknesses of the mating plates.

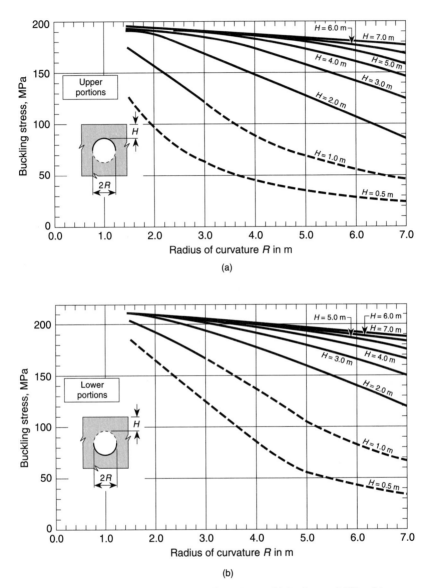

Figure 5.4 Unfactored buckling stresses for 3.0-mm-thick plates of 152 × 51 mm corrugation profile: (*a*) charts for upper portions; (*b*) charts for lower portions.

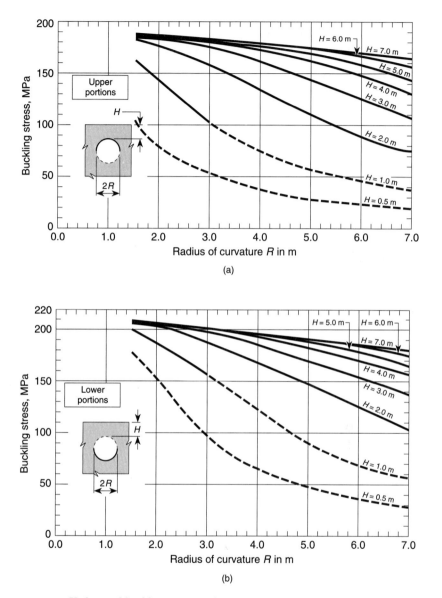

Figure 5.5 Unfactored buckling stresses for 4.0-mm-thick plates of 152 × 51 mm corrugation profile: (*a*) charts for upper portions; (*b*) charts for lower portions.

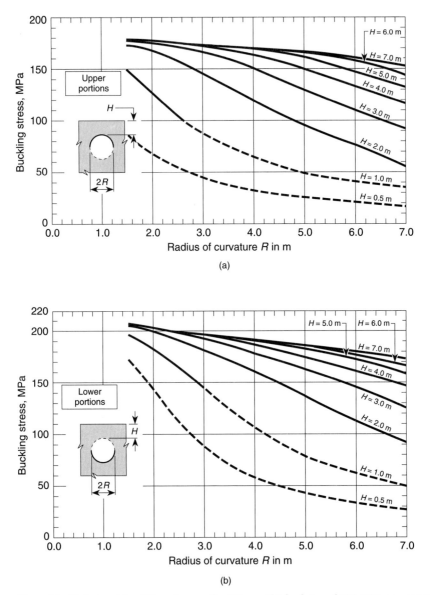

Figure 5.6 Unfactored buckling stresses for 5.0-mm-thick plates of 152 × 51 mm corrugation profile: (*a*) charts for upper portions; (*b*) charts for lower portions.

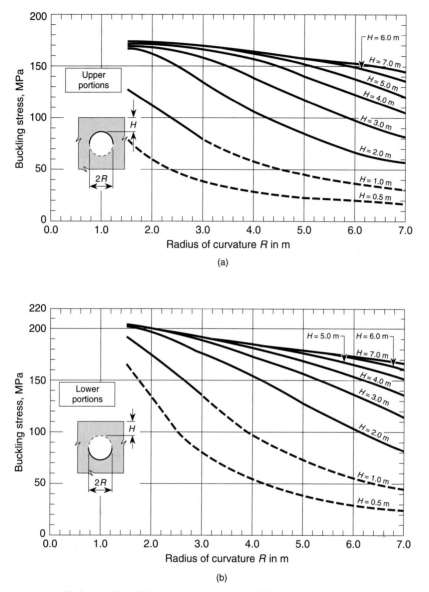

Figure 5.7 Unfactored buckling stresses for 6-mm-thick plates of 152×51 mm corrugation profile: (a) charts for upper portions; (b) charts for lower portions.

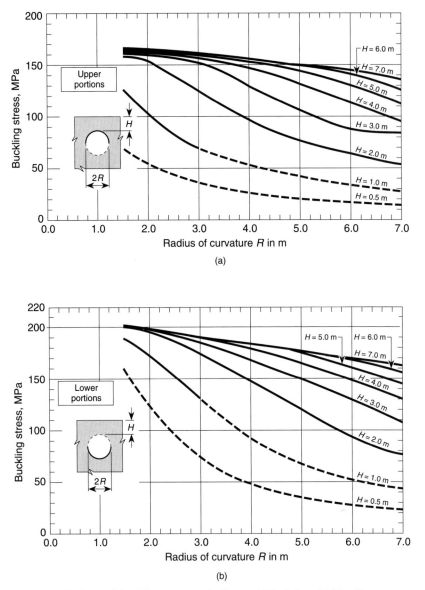

Figure 5.8 Unfactored buckling stresses for 7-mm-thick plates of 152 × 51 mm corrugation profile: (*a*) charts for upper portions; (*b*) charts for lower portions.

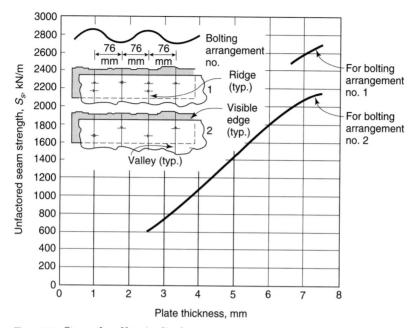

Figure 5.9   Strengths of longitudinal seams bolted with 20-mm-diameter bolts.

5.1.6   Handling stiffness

For elliptical and round pipes, the upward crown deflection during backfilling is not permitted to exceed δ_{crown} which is given by

$$\delta_{\text{crown}} = \mu \frac{D_h^2}{d} \tag{5.14}$$

where D_h is the conduit span, d the depth of the corrugation profile, being 51 mm for 152×51 mm corrugations, and μ is a factor which depends upon the span-to-rise ratio of the conduit and is read from Fig. 4.10 (Chap. 4). For structures having other than round or elliptical conduits, the upward crown deflection during backfilling is not permitted to exceed 5 percent of the conduit rise, unless it can be demonstrated that exceeding this limit will not cause the bending stresses to exceed the yield stresses which could damage the conduit wall permanently.

   In addition to the preceding requirement, the rigidity of the conduit wall is controlled for handling during construction by the recommendation that the flexibility factor of the pipe, FF, should not exceed 0.114 mm/N. The value of FF is calculated from

$$FF = \frac{D^2}{E\,I} \tag{5.15}$$

in which D is the equivalent pipe diameter, being equal to the pipe perimeter divided by π; and in which the units of the variables are compatible to give FF in mm/N. For elliptical pipes,

$$D \simeq 0.5(D_h + D_v) \qquad (5.16)$$

It is noted that the limit of FF given here may be exceeded provided that it can be demonstrated that, due to improved construction techniques, the pipe can be assembled safely even if it has a higher value of FF.

Equation (5.15) is based on the assumption that the thickness of the conduit wall is uniform around the circumference of the pipe. If a pipe has plates of different thicknesses, then the equivalent moment of inertia I can be calculated approximately by taking a weighted average of the moments of inertia of the various plates taken on the basis of their circumferential lengths.

5.1.7 Minimum depth of cover

It is required that the minimum depth of cover in meters should be the larger of

$$\frac{D_h}{6}\left(\frac{D_h}{D_v}\right)^{0.5} , \quad 0.4\left(\frac{D_h}{D_v}\right)^{2} \text{ and } 0.6 \text{ m}$$

As discussed in Chap. 4, Sec. 4.6, this requirement safeguards against the failure of the soil cover.

5.1.8 Extent of backfill

As will be discussed in Chap. 6, the engineered backfill is required by the OHBDC to be placed and compacted in layers not exceeding 300 mm of compacted thickness, each layer being compacted to the required density prior to the addition of the next layer. Such an engineered backfill must extend transversely at least one-half of the conduit span on each side of the conduit, and vertically up to the minimum depth of cover specified immediately preceding. The OHBDC requirements for the minimum extent of the engineered backfill are illustrated in Fig. 5.10.

5.1.9 Differences in radii of curvature

The OHBDC requires that the radius of curvature of the conduit at any location away from the conduit wall should not be less than one-fifth of the radius of curvature at the crown. This requirement guards against the undesirable situation of having excessive pressure on the

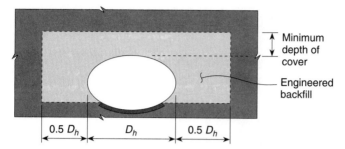

Figure 5.10   Minimum extent of engineered backfill.

soil, which is usually approximated as the pressure at the crown multiplied by the ratio of the crown radius and the local radius. It is important that the ratio of the radii of the bottom and haunch portions do not exceed the square of the coefficient of passive soil reaction. This is to avoid shear failure of the soil where the theoretical soil pressure P_h at the haunch changes to P_b under the invert as shown in Fig. 4.7 (Chap. 4).

5.1.10   Differences in plate thicknesses

The thicknesses of the plates meeting at a longitudinal seam are required to differ by not more than 1.0 mm if the thinner of the mating plates has a thickness of less than 3.1 mm; and not more than 1.5 mm if the thinner of the mating plates has a thickness between 3.1 and 3.5 mm. When each plate thickness exceeds 3.5 mm, there is no restriction on the difference in plate thicknesses at a longitudinal seam.

5.1.11   Example 5.1

The conduit wall of a soil-steel bridge with a horizontally elliptical conduit is designed in this example by the OHBDC method. The cross section of the bridge is shown in Fig. 5.11. The depth of cover and the radii of curvature of the conduit comply with the requirements of the code. The backfill for the structure is chosen as granular A soil compacted to a minimum of 90 percent standard Proctor density; for this soil, which falls into group I, Table 5.1 gives $E' = 13.0$ MPa. Other relevant details of the bridge are as follows:

Span, $D_h = 6.77$ m

Rise, $D_v = 4.88$ m

Top radius, $R_c = 4.57$ m

Side radius, $R_s = 2.05$ m

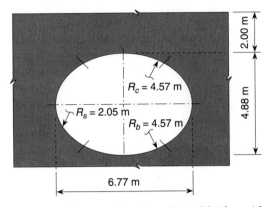

Figure 5.11 Cross section of a soil-steel bridge with horizontally elliptical conduit used in Examples 5.1, 5.3, and 5.5.

Bottom radius, $R_b = 4.57$ m

Depth of cover, $H = 2.00$ m

Modulus of elasticity of steel, $E = 2.0 \times 10^5$ MPa

Yield stress of steel, $f_y = 230$ MPa

Unit weight of soil, $\gamma = 21$ kN/m^3

Effective secant modulus of soil, $E_s^* = 25$ MPa

It is required to design the pipe of this structure from corrugated steel plates having 152×51 mm corrugation profile and having thicknesses that are generally available in Canada which are identified in Table 1.3 (Chap. 1).

From the preceding, $D_h/D_v = 1.39$ and $H/D_h = 0.29$; then, for these ratios, Fig. 4.22 in Chap. 4 gives $A_f = 1.12$, and W, calculated according to Fig. 4.1, is found to be 401 kN/m.

Using the values of A given in Chap. 1, Table 1.3 and the preceding data, the values of C_s are found from Eq. (5.4) to be 0.15, 0.11, 0.09, 0.07, and 0.06 for nominal plate thicknesses of 3.0, 4.0, 5.0, 6.0, and 7.0 mm, respectively; for these values of C_s and the values of W, and A_f obtained in the foregoing, Eq. (5.3) gives $T_D = 191$, 200, 204, 209, and 211 kN/m, respectively, for the aforementioned plate thicknesses. It can be seen that for the case under consideration, C_s has a relatively insignificant effect on the dead-load thrust. It is convenient for preliminary design to use the highest value of T_D, i.e., 211 kN/m.

Live-load thrust T_L is calculated by using Eq. (5.5) where the value of $\sigma_m m_f (1 + DLA)$ can be read directly from Fig. 5.2 to be 14.1 kPa for $H = 2.0$ m. The value of l_t, which is defined in Fig. 4.24 (Chap. 4), is 1.20

$+ (2 \times 2.0) = 5.20$ m. Since l_t is smaller than D_h, the former is used to calculate T_L, so that

$$T_L = 0.5 \times 5.20 \times 14.1$$

$$= 37 \text{ kN/m}$$

Since α_D and α_L are 1.25 and 1.40, respectively, the conduit wall thrust due to factored loads, $T_f = (1.25 \times 211) + (1.40 \times 37) = 316$ kN/m.

The compressive failure stress, f_b, for the upper zone is calculated first. The reduction factor ρ is found from Eq. (5.9) to be 2.0/4.57, i.e., 0.662. The modified modulus of soil reaction is obtained from Eq. (5.11):

$$E_m = 13.0 \left\{ 1 - \left(\frac{4.57}{4.57 + 2.00} \right)^2 \right\}$$

$$= 6.71 \text{ MPa}$$

The remaining calculations for buckling analysis depend upon the properties of the corrugated plate.

To start with, we select the 3.0-mm-thick plate for which $I = 1057.3$ mm^4/mm and $r = 17.33$ mm, as can be seen from Table 1.3 in Chap. 1. Then λ is calculated for this plate from Eq. (5.12) as

$$\lambda = 1.22 \left\{ 1.0 + 1.6 \left(\frac{200{,}000 \times 1057.3}{6.71 \times 4570^3} \right)^{0.25} \right\}$$

$$= 1.483$$

It is noted that R_c in the preceding calculation is used in mm for the units to be self-consistent.

From Eq. (5.10)

$$K = 1.483 \left\{ \frac{200{,}000 \times 1057.3}{6.71 \times 4570^3} \right\}^{0.25}$$

$$= 0.199$$

To see which of the two equations for f_b is applicable, R_e is calculated from Eq. (5.8).

$$R_e = \frac{0.01733}{0.199} \left\{ \frac{6 \times 200{,}000 \times 0.468}{230} \right\}^{0.50}$$

$$= 5.12 \text{ m}$$

It can be seen that, in this calculation, r is used in meters in order to get R_e also in meters.

Since the radius of curvature of the upper zone, i.e., 4.57 m, is less than R_e, the elastic-plastic equation for buckling, i.e., Eq. (5.7), applies, so that

$$f_b = 0.8 \times 1.0 \left[230 - \frac{(230)^2}{12 \times 200000 \times 0.662} \left(\frac{0.199 \times 4570}{17.33} \right)^2 \right]$$

$$= 110.6 \text{ MPa}$$

Accordingly, the factored strength of the conduit wall to sustain thrust, i.e., ϕR_n, is calculated as follows:

$$\phi R_n = 110.6 \times 3.522 = 389 \text{ kN/m}$$

Since this value of factored strength is larger that the thrust due to factored loads, T_f, (=316 kN/m), a thickness of 3 mm is more than adequate to satisfy the strength requirements for the upper segment of the conduit wall. Because the margin between the factored strength and thrust due to factored load is much larger than between the thrusts for plate thicknesses 7.0 and 3.0 mm, there is no need to repeat the calculations for T_D.

It may be noted that in the metric system of Canada, the minimum thickness of the plate with 152×51 mm (6×2 in) corrugation profile is 3 mm; the corresponding minimum thickness available in the U.S. customary system of units is 2.7 mm (0.106 in).

The buckling stresses for the side segments which have $R = 2.05$ m, and for the bottom segments for which $R = 4.57$ m, can be similarly calculated. As expected, f_b for bottom segments is higher than f_b for top segments, and f_b for side segments is higher still. These segments can, therefore, be of 3.0-mm plate unless a greater thickness is required by considerations other than those of strength requirements.

The nominal seam strength for 3.0 mm plates, read from Fig. 5.9 for bolting arrangement 1, is 760 kN/m. Corresponding to $\phi = 0.70$, the factored seam strength is 532 kN/m. Since the thrust due to factored loads is only 318 kN/m it is obvious that, from the standpoint of seam strength, the 3.0-mm plate will suffice.

From Eq. (5.16), the equivalent diameter $D = 0.5(6.77 + 4.88) = 5.83$ m. Using this value of D, and $I = 1057.3$ mm^4/mm for the 3.0 mm plate, Eq. (5.15) gives

$$FF = \frac{5830^2}{200,000 \times 1057.3}$$

$$= 0.161 \text{ mm/N}$$

Since this value of FF is larger than the permitted value of 0.114, the 3.0-mm plate is not favored, as it will be difficult to construct the pipe from it. A 4.0-mm-thick plate is now tried for which $I = 1457.6$ mm⁴/mm; for the pipe with this plate, $FF = 0.117$. This value is marginally larger than 0.114; however, the 4.0-mm plate may be accepted as being close enough to the limit which is nevertheless arbitrary.

5.1.12   Example 5.2

A pipe-arch having the details as shown in Fig. 5.12 is designed in this example by the OHBDC method. It will be noted that this figure gives not only the radii of curvature of the various segments, but also their subtended angles. The backfill chosen for this structure is the same as that for Example 5.1, for which $E' = 13.0$ MPa. Other relevant details of the bridge are as follows:

span, $D_h = 9.35$ m

Rise = 6.30 m

$D_v \ (=2 \times 4.27) = 8.54$ m

Top radius, $R_c = 4.73$ m

Side radius, $R_s = 1.58$ m

Bottom radius, $R_b = 10.59$ m

Depth of cover, $H = 1.83$ m

Modulus of elasticity of steel, $E = 2.0 \times 10^5$ MPa

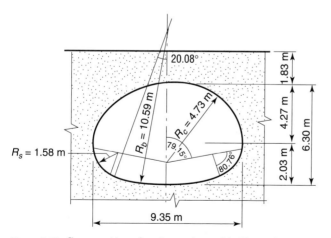

Figure 5.12 Cross section of a pipe-arch used in Examples 5.2, 5.4, and 5.6.

Yield stress of steel, $f_y = 230$ MPa

Unit weight of soil, $\gamma = 21$ kN/m³

Effective secant modulus of soil, $E_s^* = 25$ MPa

As for Example 5.1, the purpose of this exercise is to design the conduit wall with 152×51 mm corrugated plates available in Canada. The properties for these plates are given in Chap. 1, Table 1.3.

From the preceding information and $D_h/D_v = 1.09$ and $H/D_h = 0.20$; for these ratios Fig. 4.22 (Chap. 4) gives $A_f = 1.20$. From Fig. 4.1 the value of W is found to be 545 kN/m. By using Eq. (5.4), the values of C_s for plate thickness of 3.0, 4.0, 5.0, 6.0, and 7.0 mm are found to be 0.26, 0.19, 0.16, 0.12, and 0.11, respectively; for these values of C_s, the values of T_D obtained by using Eq. (5.3) are 318, 321, 322, 323, and 323, kN/m, respectively. As for Example 5.1, the highest thrust, i.e., 323 kN/m, is used for preliminary calculations. Using the same procedure as used in Example 5.1, the live-load thrust T_L is found to be 39 kN/m.

For α_D and $\alpha_L = 1.25$ and 1.40, respectively, the thrust T_f due to factored loads $= (1.25 \times 323) + (1.40 \times 39) = 458$ kN/m.

The compressive failure stress, f_b, of the upper zone is calculated first. The reduction factor ρ is then calculated from

$$\rho = \sqrt{1.83/4.73} = 0.622$$

The modified modulus of soil reaction E_m is found from Eq. (5.11):

$$E_m = 13.0\left\{1 - \left(\frac{4.73}{4.73 + 1.83}\right)^2\right\}$$

$$= 6.24\text{MPa}$$

A thickness of 5 mm is selected for the top segment. Using the properties of a plate with this thickness, obtained from Chap. 1, Table 1.3, Eqs. (5.8) and (5.10), respectively, give $R_e = 4.232$ m and $K = 0.23$. In this case, Eq. (5.6) is applicable; it gives $f_b = 73.66$ MPa. The area of cross section of the 5-mm-thick corrugated plate is 6.149 mm²/mm. Accordingly, the factored compressive strength of the plate $= 73.66 \times 6.149 = 453$ kN/m. Since the factored strength is very close to thrust due to factored loads, being 458 kN/m, the 5-mm-thick plate will suffice for the top segment.

It can similarly be found that a plate thickness of 4.0 mm will be sufficient for the side and bottom segments.

The nominal seam strength of 4-mm-thick plate for bolting arrangement 1 is found to be 1100 kN/m from fig. 5.9. Hence the factored seam strength $= 0.7 \times 1100 = 770$ kN/m. Since this factored strength is larger than the thrust due to factored loads, i.e., 458 kN/m, bolting arrangement 1 is suitable. As noted in Fig. 5.9, this arrangement incorporates bolts with a diameter of 20 mm.

For the calculation of the stiffness factor FF, the equivalent diameter D is calculated by using the subtended angles and radii of curvature given in Fig. 5.12:

$$D = \left(\frac{2}{\pi}\right)(4.73 \times 79.15 + 1.58 \times 80.76 + 10.59 \times 20.08)\left(\frac{\pi}{180}\right)$$

$$= 7.94 \text{ m}$$

The moments of inertia of 4- and 5-mm-thick plates are 1458 mm^4/mn and 1867 mm^4/mm, respectively. The average of these moments of inertia weighted on the basis of their reference lengths is given by

$$I = \frac{1867 \times \pi \times 79.15 + 1458 \times \pi \times 80.76 + 1458 \times \pi \times 20.08}{\pi \times 79.15 + \pi \times 80.76 + \pi \times 20.08}$$

$$= 1639 \text{ mm}^4/\text{mm}$$

Using this value of I, FF is calculated from

$$FF = \frac{7440^2}{2.0 \times 10^5 \times 1867.1} = 0.148 \text{ mm/N}$$

The value of the flexibility factor obtained is larger than the permitted limit of 0.114 mm/N. The designer can either specify measures, such as temporary props and ties, to be used during construction or, alternatively, the thickness of the conduit wall can be increased to meet the flexibility criterion which, as mentioned earlier, is not binding. It will be found that the flexibility criterion is satisfied by a plate thickness of 7.0 mm (0.276 in).

5.1.13   Design tables for circular and elliptical pipes

Tables 5.3 and 5.4 provide the conduit wall thicknesses that meet the design criteria of the OHBDC (Ref. 4) for round pipes and horizontally elliptical pipes with a span-to-rise ratio $= 1.7$ each having structural backfill comprising group I soils compacted to a minimum of 90 percent standard Proctor density. It can be seen in these tables that the thickness of the conduit wall is often governed by the flexibility criterion which, as noted previously, can be violated if adequate measures are taken to handle the pipe during construction.

TABLE 5.3   Design Table for Round Pipes Designed by the OHBDC

D_h, m	Conduit wall thickness to satisfy flexibility criterion	Segment (see note)																	
			Conduit wall thickness to satisfy the design requirements for the following values of H, mm															
			0.6	0.8	1.0	1.5	2.0	2.5	3.0	4.0	5.0	6.0	8.0	10.0	12.5	15.0	17.5	20.0	25.0
3.0	3.0 mm	lower													4.0				
		upper															6.0		
3.5	3.0 mm	lower														5.0			
		upper															6.0	7.0	
4.0	3.0 mm	lower							3.0										
		upper														5.0	6.0	7.0	
4.5	3.0 mm	lower																	
		upper														7.0			
5.0	4.0 mm	lower													5.0	6.0			
		upper																	
5.5	4.0 mm	lower																	
		upper													7.0				
6.0	5.0 mm	lower												5.0	6.0				
		upper																	
6.5	5.0 mm	lower																	
		upper												6.0	7.0				
7.0	6.0 mm	lower																	
		upper												6.0					
7.5	7.0 mm	lower																	
		upper											7.0						
8.0	No thickness satisfies the flexibility criterion	lower																	
		upper											6.0	7.0					
8.5		lower						5.0		4.0		5.0	6.0						
		upper																	
9.0		lower																	
		upper				6.0	5.0					6.0	7.0						

Depth of cover unsuitable (lower-left region). *No thickness satisfies the design requirements* (lower-right region).

NOTE: "Lower segment" includes side and bottom segments of the conduit wall.

5.2   AASHTO Method

The American Association of State Highway and Transportation Officials (AASHTO) specifications (Ref. 1) implicitly divide soil-steel bridges into two groups, i.e., structures with and without acceptable special features which are meant to enhance their behavior and performance. The acceptable features include:

1. continuous longitudinal stiffeners of either steel or reinforced concrete, attached to the pipe at each side of the top arc

2. frequently spaced transverse stiffeners attached to the top arc

These special features are shown conceptually in Fig. 5.13.

TABLE 5.4   Design Table for Horizontal Elliptical Pipes with Span/Rise Ratio = 1.7, Designed by the OHBDC

D_h, m	Conduit wall thickness to satisfy flexibility criterion	Segment (see note)	Conduit wall thickness to satisfy the design requirements for the following values of H, mm																
			0.6	0.8	1.0	1.5	2.0	2.5	3.0	4.0	5.0	6.0	8.0	10.0	12.5	15.0	17.5	20.0	25.0
3.0	3.0 mm	lower / upper														4.0	4.0	5.0	6.0
3.5	3.0 mm	lower / upper															5.0 / 6.0		7.0
4.0	3.0 mm	lower / upper													4.0		5.0	6.0	
4.5	3.0 mm	lower / upper							3.0								7.0	
5.0	3.0 mm	lower / upper													4.0	5.0	6.0 / 7.0		
5.5	3.0 mm	lower / upper																	
6.0	3.0 mm	lower / upper																	
6.5	4.0 mm	lower / upper														7.0		
7.0	4.0 mm	lower / upper												4.0 / 5.0	6.0			
7.5	5.0 mm	lower / upper													7.0			
8.0	5.0 mm	lower / upper																
8.5	6.0 mm	lower / upper										6.0						
9.0	6.0 mm	lower / upper								4.0	5.0 / 5.0	7.0						

Depth of cover unsuitable

No thickness satisfies the design requirements

NOTE: "Lower segment" includes side and bottom segments of the conduit wall.

As discussed later in the section, the structural design of soil-steel bridges with special features entails little in the way of calculation; it essentially involves the selection of the conduit wall thickness from a table.

While the AASHTO specifications permit both the working-stress and load-factor design method, only the latter is dealt with herein.

Despite the fact that the current edition of the AASHTO specifications uses only the U.S. customary units in the main sections, both systems of units are used in this section. This is done mainly for the sake of wider appeal and to facilitate comparison of the outcome of designs by different methods.

Some of the AASHTO requirements for the design of soil-steel bridges are not explicit enough, and thus require interpretation. These

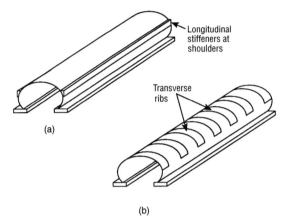

Figure 5.13 Special features acceptable to AASHTO specifications: (a) longitudinal stiffeners; (b) transverse stiffeners.

and other AASHTO requirements have been interpreted and presented in what follows in a format which is consistent with the format of the OHBDC method presented in Sec. 5.1.

5.2.1 Design criteria

Similarly to the OHBDC, the AASHTO structural design of the conduit wall is based upon the neglect of bending moments and the consideration of only the dead and live loads. The conduit wall is designed against three possible modes of failure, namely:

1. failure by crushing

2. failure by buckling

3. failure of longitudinal seams

The AASHTO structural criterion corresponding to the Load Factor Design (LFD) method can be expressed by

$$\phi\, R_n \geq \alpha_L L(1 + DLA) + \alpha_D D$$

where ϕ = the capacity modification factor, being 0.67
  R_n = the nominal capacity of the conduit wall corresponding to the failure mode under consideration
  α_L = the live-load factor, being effectively 2.171
  α_D = the load factor corresponding to the dead load of soil, being effectively 1.95
  L = the thrust due to live loads
  DLA = the impact factor
  D = the thrust due to the dead load of soil

It is noted that the conduit wall thrust due to factored loads corresponding to the load factor design is denoted as T_l in the AASHTO specifications, using which

$$T_l = \alpha_L L(1 + DLA) + \alpha_D D \tag{5.17}$$

5.2.2 Conduit wall thrust

The factored conduit wall thrust T_l is calculated by the ring compression theory according to which

$$T_l = 0.5\{\alpha_L(1 + DLA)P_L + \alpha_D P_D\}D_h \tag{5.18}$$

where P_L and P_D are the nominal equivalent uniformly distributed pressures at the crown level due to the soil dead load and live load, respectively.

The pressure P_D is simply the free-field overburden pressure at the crown. In the absence of a wearing course,

$$P_D = \gamma H \tag{5.19}$$

where γ is the soil density and H the depth of cover. When there is pavement on the embankment, its weight per unit area should be added to γH to obtain P_D.

The equivalent live-load pressure P_L is calculated by assuming that a wheel load is uniformly distributed over a square area, each side of which is 1.75 times H (Fig. 5.14). When such areas due to several wheel loads overlap, the total load is assumed to be distributed uniformly over the area contained by the outer limits of the individual areas. It is noted that this procedure is applicable only when $H > 0.61$ m (2 ft).

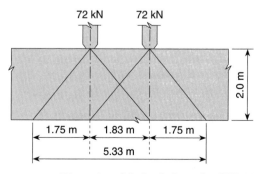

Figure 5.14 Dispersion of the load of an axle of HS-20 vehicle through fill in the longitudinal direction of the conduit.

Impact, or dynamic load allowance *DLA*, is specified to be 0.0 when the depth of cover H is more than 0.91 m (3 ft); for smaller values of H

$$DLA = \frac{50}{D_h + 125} \tag{5.20}$$

where the conduit span D_h is in feet.

5.2.3 Crushing stress

For safeguarding against the failure of the conduit wall by crushing, it is required that

$$\phi\, A\, f_y \geq T_l \tag{5.21}$$

where $\phi = 0.67$, A is the cross-sectional area of the conduit wall per unit length, and f_y is the specified minimum yield stress of the conduit wall steel.

5.2.4 Buckling stress

The conduit wall is also required to be checked for failure by buckling according to

$$\phi\, A\, f_b \geq T_l \tag{5.22}$$

where f_b is the buckling stress given by the relevant of the following two equations, and other notation is as defined earlier.

$$f_b = \frac{12E}{(K\,D_h/r)^2} \qquad \text{for } D_h \geq \overline{D}_h \tag{5.23}$$

$$f_b = f_u - \frac{f_u^2}{48E}\left(\frac{K\,D_h}{r}\right)^2 \qquad \text{for } D_h < \overline{D}_h \tag{5.24}$$

where

$$\overline{D}_h = \frac{r}{K}\left\{\frac{24E}{f_u}\right\}^{0.5} \tag{5.25}$$

f_u = specified minimum ultimate stress of the conduit wall steel
K = soil stiffness factor whose value is specified to be 0.22
D_h = the span of the conduit
E = the modulus of elasticity of conduit wall steel
r = the radius of gyration of the corrugation profile

It is recalled that $\phi = 0.67$.

5.2.5 Seam strength

The longitudinal seams are required to have strength S_s to satisfy

$$\phi S_s \geq T_l \tag{5.26}$$

where ϕ is again equal to 0.67. The values of S_s are listed in Table 4.1 (Chap. 4) for various bolting arrangements and plate thicknesses. It is noted that these values are very similar to the values of S_s specified in the OHBDC and plotted in Fig. 5.9. The bolting arrangements with two and three bolts per pitch are shown as bolting arrangements nos. 1 and 2, respectively, in this figure.

5.2.6 Handling stiffness

The AASHTO specifications recommend that the value of the flexibility factor FF should generally not exceed 0.171 mm/N (0.03 in/lb) for arches and pipe-arches, and 0.114 mm/N (0.02 in/lb) for other structures. FF is given by

$$FF = \frac{D_h^2}{EI} \tag{5.27}$$

in which I is the moment of inertia of the conduit wall per unit length and other notation is as defined earlier. Equation (5.27) is similar to Eq. (5.15) specified by the OHBDC, with the main difference that for noncircular conduits, the latter specifies D as the equivalent diameter of the pipe being equal to its perimeter divided by π. In the former equation, the actual span is used instead of D.

5.2.7 Backfill

The engineered backfill for structures with depth of cover up to 3.66 m (12 ft) is required to conform to soil classifications A-1, A-3, A-2-4, and A-2-5, specified in Table 7.5. However, for structures with higher depth of cover, the backfill is limited to conform with soil classifications A-1 and A-3. All the engineered backfill is required to be compacted to a minimum of 90 percent standard Proctor, or AASHTO T180, density.

The AASHTO specifications are not very clear on defining the required extent of the structural backfill, stating that for "ordinary installations" the backfill should extend 1.83 m (6 ft) on either side of the conduit and 0.61 to 1.22 m (2 to 4 ft) above it.

5.2.8 Structures with and without special features

Soil-steel bridges without special features are referred to in the AASHTO specifications as "structural plate pipe structures"; these structures are required to comply with all the requirements given previously in Sec. 5.2. In addition, the depth of cover for them is required to be the larger of 0.3 m (12 in) and $D_h/8$. It is noted that for this restriction the thickness of the pavement is not included in the depth of cover unless the pavement is rigid.

Those soil-steel bridges which, because of their large spans, cannot comply with the buckling and flexibility requirements given earlier in this section and which are strengthened by the acceptable features identified in Fig. 5.13, are referred to as "long-span structural plate structures." The conduit walls of these structures are exempted from complying with buckling and flexibility requirements. Instead, the top portions of the pipe are required to have the minimum thickness listed in Table 5.5 corresponding to the radius of curvature, and the bottom portions of the pipe are required to be designed against failure by crushing. The seam strength requirements given earlier also apply to both the top and bottom portions of the pipe.

The minimum specified depth of cover for long-span structural plate structures depends upon the thickness of the top portion of the pipe and its radius of curvature, and is as given in Table 5.6. It is noted that, in the case of conduit walls with transverse stiffeners, the plate thickness for use of this table can be increased provided that the moment of inertia corresponding to the plate of this increased thickness is less than or equal to the moment of inertia of the actual stiffened plate.

For long-span structural plate structures, the ratio of the radii of the top and side segments must be neither less than 2 nor greater than 5.

5.2.9   Example 5.3

The structure designed in Example 5.1 by the OHBDC method is designed in this example by the AASHTO specifications for HS20 loading. The cross section of the structure is shown in Fig. 5.11, and other basic information necessary for design calculations is repeated here in both systems of units:

TABLE 5.5   Minimum Thickness of the Top Arc of the Pipe of "Long-Span Structural Plate Structures," required by AASHTO Specifications

R_c in m (ft)	4.572 (15)	4.572–5.182 (15–17)	5.182–6.096 (17–20)	6.096–7.010 (20–23)	7.010–7.620 (23–25)
Minimum conduit wall thickness, mm (in)	2.77 (0.109)	3.51 (0.138)	4.27 (0.168)	5.54 (0.218)	6.32 (0.249)

TABLE 5.6   Minimum Depth of Cover for "Long-Span Structural Plate Structures," required by AASHTO Specifications

Conduit wall thickness of top arc mm (in)	Minimum depth of cover in m (ft) corresponding to R_t in m (ft) =				
	4.572 (15)	4.572–5.182 (15–17)	5.182–6.096 (17–20)	6.096–7.011 (20–23)	7.010–7.620 (23–25)
2.77 (0.109)	2.77 (2.5)	0.762			
3.51 (0.138)	0.762 (2.5)	0.914 (3.0)			
4.27 (0.168)	0.762 (2.5)	0.914 (3.0)	0.914 (3.0)		
4.78 (0.188)	0.761 (2.5)	0.914 (3.0)	0.914 (3.0)		
5.54 (0.218)	0.610 (2.0)	0.762 (2.5)	0.762 (2.5)	0.914 (3.0)	
6.32 (0.249)	0.610 (2.0)	0.610 (2.0)	0.762 (2.5)	0.914 (3.0)	1.219 (4.0)
7.11 (0.280)	0.610 (2.0)	0.610 (2.0)	0.762 (2.5)	0.914 (3.0)	1.219 (4.0)

Span, D_h = 6.77 m (22.21 ft)

Depth of cover, H = 2.00 m (6.56 ft)

Modulus of elasticity of steel, $E = 2.0 \times 10^5$ MPa (29×10^6 psi)

Yield stress of steel, f_y = 230 MPa (33.4 ksi)

Ultimate stress of steel, f_u = 314 MPa (45.5 ksi)

Unit weight of soil, γ = 21 kN/m³ (134 lb/ft³)

It is required to design this structure for corrugated steel plates having 152 × 51 mm (6 × 2 in) corrugation profile and having thicknesses generally available in the United States which are identified in Chap. 1 Table 1.2.

For calculating the conduit wall thrust, P_D is first calculated according to Eq. (5.19)

$$P_D = 21.0 \times 2.0 = 42.0 \text{ kN/m}^2 \text{ (879 lb/ft}^2\text{)}$$

As shown in Fig. 5.14, the 144 kN (32 kip) axle of the HS-20 vehicle, shown in Fig. 4.4a, is distributed through the 2.0-m- (6.56 ft) deep fill to a length of 5.33 m (17.49 ft), measured in the longitudinal direction of the conduit. In the transverse direction, the length of the dispersed load at the crown level is 1.75 × 2.0, or 3.5 m (11.48 ft). Consequently

$$P_L = \frac{144}{5.33 \times 3.5} = 7.7 \text{ kN/m}^2 \text{ (159 lb/ft}^2\text{)}$$

Since the depth of cover is more than 0.91 m (3.0 ft), $DLA = 0$. From Eq. (5.18), the thrust due to factored loads is given by

$$T_l = 0.5\{(2.171 \times 7.7) + (1.95 \times 42.0)\} \times 6.77$$

$$= 333.8 \text{ kN/m} = 333.8 \text{ N/mm} \ (22.9 \text{ kips/ft.})$$

From Eq. (5.21), the minimum area A to satisfy the crushing stress requirement is given by

$$A = \frac{333.8}{0.67 \times 230} = 2.166 \text{ mm}^2/\text{mm} \ (0.085 \text{ in}^2/\text{in})$$

As can be seen from Table 1.2 (Chap. 1), even the thinnest plate used with 152×51 mm (6×2 in) corrugations, being 2.7 mm (0.106 in) thick, has the cross-sectional area of 3.293 mm²/mm (0.130 in²/in); hence, even this thinnest plate can satisfy the crushing-strength requirement.

The 2.7-mm- (0.106 in) thick plate is now tried for suitability against failure by buckling. From Table 1.2, for this plate $r = 17.32$ mm (0.681 in). For $K = 0.22$, Eq. (5.25) gives

$$\bar{D}_h = \frac{17.32}{0.22} \left\{ \frac{24 \times 2.0 \times 10^5}{314} \right\}^{0.5}$$

$$= 9733 \text{ mm} = 9.733 \text{ m} \ (31.93 \text{ ft})$$

Since $D_h < \bar{D}_h$, Eq. (5.24) applies for the calculation of the buckling stress, so that

$$f_b = 314 - \frac{314^2}{48 \times 2.0 \times 10^5} \left\{ \frac{0.22 \times 6770}{17.32} \right\}^2$$

$$= 238 \text{ MPa} \ (34.5 \text{ ksi}).$$

It can be seen that this buckling stress is even higher than the yield stress of steel which is the same as the "crushing stress." Therefore, it is concluded that the 2.7-mm- (0.106 in) thick plate also satisfies the buckling strength requirements.

For checking the stiffness of the pipe for handling, its flexibility FF is calculated by Eq. (5.27) for the aforementioned plate thickness, for which $I = 990.1$ mm⁴/mm (0.060 in⁴/in) as obtained from Table 1.2 (Chap. 1). Then

$$FF = \frac{6770^2}{2.0 \times 10^5 \times 990.1} = 0.23 \text{ mm/N} \ (0.04 \text{ in/lb})$$

Since this value of *FF* is much greater than the recommended value of 0.114 mm/N (0.02 in/lb), the selected plate thickness is not suitable. The minimum value of I to satisfy the handling stiffness requirements can be obtained from

$$I = \frac{6770^2}{0.114 \times 2.0 \times 10^5} = 2010 \text{ mm}^4/\text{mm (0.122 in}^4/\text{in)}$$

From Table 1.4 (Chap. 1), it can be seen that the minimum plate thickness to have I greater than this value is 5.4 mm (0.213 in).

It is concluded that for structural requirements alone, the pipe can be made of plates having a thickness of 2.7 mm (0.106 in). However, from the consideration of handling stiffness, it is desirable to use 5.4-mm (0.213 in) thick plates.

5.2.10   Example 5.4

The pipe-arch designed in Example 5.2 by the OHBDC and shown in Fig. 5.12, is redesigned in this example by the AASHTO specifications corresponding to the HS20 loading. The relevant design parameters are reproduced following in both systems of units:

Span D_h = 9.35m (30 ft 8 in)

Depth of cover, H = 1.83 m (6 ft)

Modulus of elasticity of steel, E = 2.0 × 10⁵ MPa (29,000 ksi)

Yield stress of steel, f_y = 230 MPa (33.4 ksi)

Ultimate stress of steel, f_u = 314 MPa (45.5 ksi)

Unit weight of soil, γ = 21 kN/m² (134 lb/ft³)

It is required to design the thicknesses of the 152 × 51 mm (6 × 2 in) corrugated plate for top, side, and bottom segments of the pipe-arch. Since the explanation for the various steps of calculation are already given in conjunction with Example 5.3, the steps of calculation for the example at hand are given with minimal explanation.

$$P_D = 21 \times 1.83 = 38.43 \text{ kN/m}^2(804 \text{ lb/ft}^2)$$

$$P_L = \frac{144}{4.42 \times 5.03} = 6.47 \text{ kN/m}^2(135 \text{ lb/ft}^2)$$

$$T_l = 0.5\{(2.171 \times 6.47) + (1.95 \times 38.43)\}9.35$$

$$= 416.0 \text{ kN/m}(28.5 \text{ kips/ft})$$

The minimum area A to satisfy the crushing-strength requirement (Eq. 5.21) is given by

$$A = \frac{416.0}{0.67 \times 230} = 2.70 \text{mm}^2/\text{mm}(0.106 \text{ in}^2/\text{in})$$

Even the thinnest plate having a thickness of 2.7 mm (0.106 in) and a cross-sectional area of 3.293 mm²/mm (0.130 in²/in), will satisfy the crushing-strength requirement. The buckling strength is investigated for this plate.

As calculated in Example 5.3, $\overline{D}_h = 9.733$ m (31.93 ft). Since $D_h < \overline{D}_h$, Eq. (5.24), is used for calculating f_b

$$f_b = 314 - \frac{314^2}{48 \times 2.0 \times 10^5} \left[\frac{0.22 \times 9350}{17.32} \right]^2$$

$$= 169 \text{MPa}(24.52 \text{ksi})$$

The factored buckling stress = $0.67 \times 169 = 113.23$ MPa (16.42 ksi).

The stress due to $T_l = 416.0/3.293 = 126$ MPa (18.32 ksi). Since Eq. (5.21) is not satisfied, the next available thickness, i.e., 3.51 mm (0.138 in), is tried. By repeating the preceding calculations for this thicker plate, it will be found that the 3.51-mm- (0.138 in) thick plate satisfies the requirements of buckling strength. Similarly, the requirements of seam strength, Eq. (5.26), will be satisfied by this plate incorporating even the weakest of the three bolting arrangements, i.e., no. 1.

The flexibility factor for the 3.51-mm- (0.138 in) thick plate is calculated from Eq. (5.27).

$$FF = 9350^2/(2.0 \times 10^5 \times 1280.9) = 0.341 \text{ mm/N (0.60 in/lb)}$$

Since the value of FF for 3.51-mm- (0.138 in) thick plate is much greater than the permitted maximum of 0.171 mm/N (0.03 in/lb), a thicker plate will have to be selected to satisfy the handling criterion. It will be found that only a 7.0-mm- (0.276 in) thick plate can satisfy this criterion.

5.3 Duncan Method

The Duncan method (Ref. 3), while basically derived for structures with pipes made of corrugated aluminium sheets, is also applicable for the design of the pipes of soil-steel bridges. It differs from the Ontario and AASHTO methods in that it disregards the failure of the conduit wall by buckling, and controls the design through the formation of plastic hinges when the depth of cover is shallow, and through yielding by axial forces when it is deep. However, similarly to the two former methods, the Duncan method is also based upon consideration of only live loads and dead loads, including the soil loads.

The Duncan method is based upon the Working Stress Design approach. However, the format of the method is such that it can be transformed readily to Load Factor Design or Load and Resistance Factor Design.

5.3.1 Construction stage

The structure is designed for two conditions. For the first condition, which corresponds to a stage during construction, the fill is at the crown level, i.e., $H = 0$. For this condition, the bending moments M and thrust T in the conduit wall are calculated without consideration of live loads and are given by

$$M = R_B K_{m1} \gamma D_h^3 \tag{5.28}$$

$$T = K_{p1} \gamma D_h^2 \tag{5.29}$$

where R_B, K_{m1}, and K_{p1} are calculated from Eqs. (4.8), (4.6), and (4.10), respectively, D_h is the conduit span and H is the depth of cover.

5.3.2 Completed structure

The second condition for which the structure is designed is that at which the backfill has reached the final elevation. For this condition, in which the live loads are also considered, the moment M and thrust T in the conduit wall are given by

$$M = R_B(K_{m1} \gamma D_h^3 - K_{m2} \gamma D_h^2 H) + R_L K_{m3} D_h LL \tag{5.30}$$

$$T = K_{p1} \gamma D_h^2 + K_{p2} \gamma H D_h + K_{p3} LL \tag{5.31}$$

where R_B, K_{m1}, and K_{p2} are the same as obtained previously, and K_{m2}, K_{m3}, K_{p2}, and K_{p3} are obtained from Eqs. (4.7), (4.34), (4.11), and (4.31), respectively. The values of live load LL equivalent to the AASHTO HS-20 vehicle are listed in Table 4.2 (Chap. 4) for different depths of cover.

5.3.3 Safety factor

The safety factor (SF) against the formation of plastic hinges, for the construction condition and for shallow depth of cover, is required to be a minimum of 1.65. The value of SF can be calculated from

$$SF = \frac{0.5P_p}{T} \left\{ \left[\left(\frac{M}{M_p} \right)^2 \left(\frac{P_p}{T} \right)^2 + 4 \right]^{0.5} - \left(\frac{M}{M_p} \right) \left(\frac{P_p}{T} \right) \right\} \tag{5.32}$$

where P_p is the full plastic axial force capacity of the conduit wall with no applied moments, and M_p is the full plastic moment capacity with no applied axial force.

A safety factor of 1.50 is employed for the crushing failure of the conduit wall for structures under deep covers, and also for failure of the longitudinal seams.

It is noted in Ref. 3 that if $SF = 1.65$ is used in the design to safeguard against the formation of plastic hinges, then the actual safety margin will be greater than that reflected by the assumed factor of safety, for the following three reasons.

1. Formation of a single plastic hinge, even in an arch with pinned ends, is not enough for the failure of the pipe.

2. Even after several plastic hinges have developed in the conduit wall, the whole structure does not become unstable, because of the presence of the soil support which restrains the deformations of the pipe.

3. Since steel hardens after yielding, the axial thrust and moment capacities of the conduit wall calculated by using f_y are expected to be lower-bound estimates of the actual respective capacities.

The Duncan method is based upon the assumption that the backfill of the soil-steel bridge consists of good-quality, well-graded granular material compacted adequately.

5.3.4 Example 5.5

The same structure which has been designed by the Ontario and AASHTO methods in Examples 5.1 and 5.3, respectively, is now designed by the Duncan method. The relevant basic data for this structure are given in these two examples. In addition, following Duncan's recommendations (see Chap. 7), the secant modulus of the well-compacted, well-graded backfill of granular material is taken as 4.6 MPa (100 kips/ft^2).

The ratio $H/D_h = 2.00/6.77 = 0.30$. Since this ratio is greater than 0.25, the conduit wall needs to be designed for moment and thrust only for the construction stage at which $H = 0.0$. For the completed structure, the conduit wall has to be designed only for thrust.

To start with, a nominal plate thickness of 3.4 mm is selected for which Table 1.2 (Chap. 1) gives $I = 1280.9$ mm^4/mm and $A = 4.240$ mm^2/mm. The relevant parameters required for analysis by the Duncan method are calculated (see following) corresponding to this plate thickness and other relevant data. For ready reference, the equation numbers to which the calculations correspond are given preceding the calculations.

(Eq. 4.4) $\qquad N_f = \dfrac{4.6}{200,000} \times \dfrac{6770^3}{1280.9}$

$\qquad\qquad\qquad = 5572$

(Eq. 4.6) $\qquad K_{m1} = 0.0009$

(Eq. 4.10) $\qquad K_{p1} = 0.21 \times 2.44/2.0$

$\qquad\qquad\qquad = 0.2562$

(Eq. 4.11) $\qquad K_{p2} = 0.90 - 0.50 \times 2.44/6.77$

$\qquad\qquad\qquad = 0.7198$

(Eq. 4.31) $\qquad K_{p3} = 1.25 - 0.30 = 0.95$

(Eq. 4.8) $\qquad R_B = 0.80 + 1.33 \left\{ \dfrac{2.44}{6.77} - 0.35 \right\}$

$\qquad\qquad\qquad = 0.8139$

The plastic axial capacity P_p of the conduit wall is given by

$$P_p = 4.240 \times 230$$

$$= 975.2 \text{ N/mm,} \quad \text{or kN/m} \quad (66.8 \text{ kips/ft})$$

Similarly, the plastic moment capacity M_p of the corrugated plate section is given by

$$M_p = \frac{2.30 \times 1280.9 \times 1.2}{0.5 \times 51 \times 1000}$$

$$= 13.87 \text{ kN} \cdot \text{m/m} \ (3.13 \text{ kip} \cdot \text{ft/ft})$$

It is noted that the factor 1.2 which appears in the preceding calculations is the shape factor for the corrugation profile.

The capacity of the selected conduit wall section is now checked for the construction condition of $H = 0$, for which Eqs. (5.28) and (5.29), respectively, give

$$M = 0.8139 \times 0.0009 \times 21.0 \times 6.77^3$$

$$= 4.77 \text{ kN} \cdot \text{m/m} \ (1.07 \text{ kip} \cdot \text{ft/ft})$$

$$T = 0.2562 \times 21.0 \times 6.77^2$$

$$= 246.6 \text{ kN/m} \ (16.9 \text{ kip/ft})$$

The ratios T/P_p and M/M_p are then 346.6/975.2 (=0.253) and 4.77/13.87 (=0.344), respectively. The safety factor SF for the construction stage is given by Eq. (5.31):

$$SF = \frac{0.5}{0.253} \left\{ \left[\frac{0.344^2}{0.253^2} + 4 \right]^{0.5} - \frac{0.344}{0.253} \right\}$$

$$= 2.09$$

Since this value of SF is greater than the required minimum value of 1.65, the selected section of the conduit wall is sufficient for the construction stage. The section is now checked for the completed structure for which $H = 2.0$ m (6.56 ft). For this depth of cover, Table 4.2 (Chap. 4) gives $LL = 36.5$ kN/m (2.5 kip/ft). Equation (5.31) gives

$$T = (0.2562 \times 21.0 \times 6.77^2) + (0.7198 \times 21.0 \times 2.0 \times 6.77) + (0.95 \times 36.5)$$

$$= 485.9 \text{ kN/m(33.3 kip/ft)}$$

As discussed previously, at the final stage the structure under consideration will have $H/D_h > 0.25$, because of which the soil cover can be regarded as deep and, consequently, moments need not be considered. It is clear that in the absence of bending moments, $SF = P_p/T$, as can be obtained by putting $M = 0$ in Eq. (5.31). Therefore, for the completed structure

$$SF = \frac{975.2}{485.9}$$

$$= 2.00$$

Since this safety factor is greater than the required minimum value of 1.50, the selected section of the conduit wall is considered to be sufficient for the condition of the completed structure carrying the design live load. It is noted that the next thinner plate having a nominal thickness of 2.7 mm would also have sufficed for the condition of the completed structure, but would have narrowly failed to satisfy the requirement of $SF \geq 1.65$ for the construction stage at $H = 0.0$; for this plate thickness, the thinner plate has $SF = 1.62$.

It is concluded that for the Duncan method, the structure in consideration can have the 3.4-mm-thick conduit wall for all of the top, side, and bottom segments.

5.3.5 Example 5.6

By repeating the calculations for the pipe-arch shown in Fig. 5.12 in the same manner as for Example 5.5, it will be found that a conduit wall with a minimum thickness of 5.4 mm (0.213 in) satisfies the strength criterion by the Duncan method. A wall thickness of 6.2 mm (0.244 in) is required to satisfy the handling criterion.

5.4 Miscellaneous Observations

5.4.1 Comparisons of design methods

Table 5.7 lists the design conduit wall thicknesses for the structures shown in Figs. 5.11 and 5.12 as obtained in the preceding examples from the considerations of the conduit wall strength corresponding to the completed structure, and of the conduit wall stiffness from the point of view of handling during construction. It can be seen in this table that there are significant differences between the designs resulting from the three methods. A rationalization of these differences may be found helpful in identifying the strengths and weaknesses of these methods.

The AASHTO and Duncan methods do not recognize the need for different segments of the pipe to have different design requirements. As a result of this, they lead to the same conduit wall thickness all around the pipe. In practice, the top segment of the pipe is usually either thicker than the other segments as shown (e.g., in Chap. 1, Table 1.5) or is provided with transverse stiffners which are accepted as special features by the AASHTO specifications. This practice confirms the validity of the OHBDC approach in which the strength of the top segments is calculated differently.

While the handling stiffness criteria of OHBDC and AASHTO lead to slightly different results, it should be noted that these criteria are advisory, not mandatory, in nature. The handling stiffness requirement of the Duncan method does not deal with pipe stiffness in the same fashion that the other two methods do. The OHBDC and AASHTO methods are concerned with the pipe stiffness in the earlier stages of the backfill, while the Duncan method relates to the condition when the fill has reached the crown level.

TABLE 5.7 Comparison of Conduit Wall Thicknesses Designed by Different Methods

			Thickness in mm for strength design of:			
Fig.	Example	Method	Top segment	Side segment	Bottom segment	Thickness for handling
5.11	5.1	OHBDC	3.0	3.0	3.0	4.0
	5.3	AASHTO	2.7	2.7	2.7	5.4
	5.5	Duncan	3.4	3.4	3.4	3.4
5.12	5.2	OHBDC	5.0	4.0	4.0	7.0
	5.4	AASHTO	3.5	3.5	3.5	7.0
	5.6	Duncan	5.4	5.4	5.4	6.2

All things considered, it is concluded that the OHBDC method of strength design is preferable because: (1) it accounts realistically for the strength of the top segments of the pipe and (2) it leads to an economical use of the steel around the periphery of the pipe.

As far as handling stiffness is concerned, the designer can feel free to chose either of the OHBDC and AASHTO criteria, but should specify adequate controls for restricting the pipe deformations adequately during construction.

5.4.2 CSA method

The CSA code (Ref. 2) does not specify requirements for designing soil-steel bridges. However, since this code is very similar in spirit and philosophy to the OHBDC, it is realistic to design a soil-steel bridge by using its design vehicle and load factors, but with the remaining relevant provisions being taken from the OHBDC.

5.4.3 Construction and installation

The influence of construction and installation procedures on the structural integrity of the soil-steel bridge cannot be overemphasized. The only control that the designer may have on these procedures is by specifying them on the construction drawings. Such specifications, which can be drawn from Chap. 6, might typically be given under the following headings.

- Preparation of the bedding
- Assembly of the pipe
- Dimensional check of the pipe
- Backfilling material
- Spreading of the backfill
- Compaction of the backfill
- Compaction above shoulders
- Observing the vertical displacement at crown

References

1. American Association of State Highway and Transportation Officials (AASHTO), *Standard Specifications for Highway Bridges,* Washington, D.C., 1989.
2. Canadian Standards Association (CSA), *CAN/CSA-S6-88,* "Design of Highway Bridges," Rexdale, Ontario, Canada, 1988.
3. Duncan, J. M., "Behaviour and Design of Long-Span Metal Culverts," *ASCE Journal of the Geotechnical Division,* 105(GT3), 1979, pp. 399–417.
4. Ministry of Transportation of Ontario, *Ontario Highway Bridge Design Code* (OHBDC), Downsview, Ontario, Canada, 1992.

6

Construction

John O'Brien

*Alberta Department of Transportation
and Utilities
Airdrie, Alberta, Canada*

Baidar Bakht

*Ministry of Transportation of Ontario
Downsview, Ontario, Canada*

Cam Mirza

*Strata Engineering Corporation
Don Mills, Ontario, Canada*

Construction procedures have a profound influence on the integrity of soil-steel bridges. Somewhat differently from other types of bridges, deficiencies in the construction of a soil-steel bridge may manifest themselves not only during or immediately after construction, but a few years after the construction as well. It is now well known that a soil-steel bridge may be able to accommodate minor errors in design; however, any deficiencies in the construction process can lead to catastrophic consequences. It cannot be overemphasized that soil-steel bridges should be constructed under strict and competent supervision, and with well-established, sound construction procedures.

Over a number of years, engineers, contractors, and the corrugated steel pipe industry in North America have evolved particular procedures to construct soil-steel bridges. These procedures are presented in this chapter.

6.1 Site Preparation

Site preparation and preconstruction planning are essential for ensuring that the job is well organized and consequently free from problems.

Besides excavation of the foundation and its dewatering, which are dealt with later in the chapter, the various factors which should be considered in the preparation of the site are presented in the following under separate headings.

6.1.1 Storage and subassembly area

Corrugated metal plates for the entire pipe are delivered to the site, nested in bundles, examples of which can be seen in Fig. 6.1. A leveled area large enough for the purpose should be set aside for storing these plates, which are preferably presorted while being unloaded. This storage area is also required for storing nuts and bolts and, if necessary, for preassembling the segments of the pipe.

A grassy and well-compacted area, such as a pasture, is preferable for the storage and preassembly area, as it does not hinder the movement of construction equipment and personnel in wet weather. The storage area should ideally be within about 200 m (220 yd) of the location of the soil-steel bridge; however, even a distance of up to 1 km (0.6 mi) may prove acceptable in certain circumstances.

A typical preassembly and storage area is shown conceptually in Fig. 6.2, in which it can be seen that the plates, nuts, and bolts are stockpiled close to the access road to the area. If preassembly of plate segments away from the bridge location is required, then the area must also have room, not only for the assembly of one or more segments

Figure 6.1 Stacks of curved corrugated plates.

simultaneously, but also for the crane which transports the plates from the stockpiles to the assembly area. If the structure is large, it may be preferable to have more than one area set aside for subassembly.

6.1.2 Access roads

The foundations of soil-steel bridges are usually at a much lower elevation than the rest of the ground, as a result of which the access roads to the construction site can be quite steep. Since steep and mud-slicked roads can be difficult to maneuver on, it is advisable to dress them with gravel and also to limit the steepness of the grades as much as possible.

The access road should ideally be merged with the bedding of the pipe at one end. If the bridge conveys water, the access road should be merged with the downstream end of the bedding of the pipe.

6.1.3 Working platform

A working platform, from which a crane can operate parallel to the axis of the pipe, can speed up the erection of the pipe considerably. Clearly, such a platform should be level and be such as to permit the free movement of the crane along the entire length of the pipe; the crane is required for hauling the individual plates into place for the erection of the pipe.

A preferable layout of the construction site for a soil-steel bridge crossing a stream is shown in Fig. 6.3. This figure also shows that the

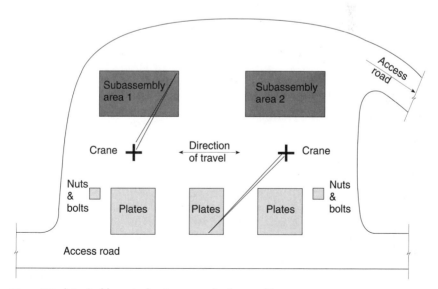

Figure 6.2 A typical layout of a storage and subassembly area.

stream is diverted to one side of the pipe, and the area to facilitate the erection of the pipe is set up on the other side. It should be noted that the access road is extended to the downstream end of the pipe so that a vehicle can be driven to the inside of the pipe if necessary.

It is obvious that the site plan for every soil-steel bridge cannot be exactly as shown in Fig. 6.3. However, if the general scheme of the layout is similar to the one shown, then the erection of the pipe can be expected to be greatly facilitated.

6.2   Excavation and Dewatering

6.2.1   Excavation

In nearly all cases, it is either necessary or at least advisable to excavate the existing ground in preparation for the construction of a soil-steel bridge. While the actual extent of the excavation depends upon the quality of the foundation and upon design considerations, the following examples will help to illustrate the preferred practice.

When the ground is more or less at the invert level, the transverse profile of the excavation may be as shown in Fig. 6.4a. It can be seen in this figure that the horizontal portion of the excavated ground is about 0.6 m (2 ft) below the invert level, and extends laterally to about 1.5 m (5 ft) on either side of the conduit. Beyond the horizontal portion, the

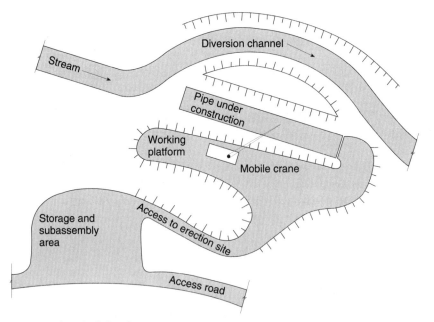

Figure 6.3 A typical site plan.

ground is excavated laterally to an inclination of about 1:1. Figure 6.4*a* also shows the cross section of the temporary diversion channel which would be required if the proposed structure were to be constructed over a water course.

Figure 6.4*b* shows the cross section of the ground and the proposed conduit for the case when the ground level is well above the invert level. In this case, also, the horizontal portion of the excavated ground is about 0.6 m (2 ft) below the invert, and extends to about 1.5 m (5 ft) laterally on either side of the conduit. There is practically no difference between the excavated ground profiles for the two cases illustrated in Fig. 6.4*a, b*.

When the ground line is above the crown of the conduit, the excavation may become quite deep; in such cases (as shown in Fig. 6.4*c*), it

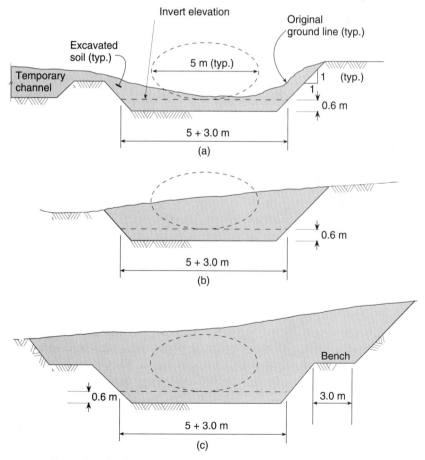

Figure 6.4 Examples of schemes of excavation: (*a*) existing ground at invert level; (*b*) existing ground above invert level; (*c*) existing ground above crown.

may be advisable to provide about 3.0-m- (10 ft) wide benches. As can be seen in this figure, the width and elevation of the horizontal portion of the excavation for the case under consideration are the same as those for the previous cases. However, under certain circumstances it may be preferable to extend the width of this horizontal portion so as to provide a platform for the preassembly of the conduit walls; this is discussed later in the chapter. The extended width of the horizontal portion also provides a level surface for movement of the crane during construction. It is noted that the stability of the bench must be checked for the crane loads before permitting the crane to be used.

6.2.2   Rock ledges

A rock ledge in the ground directly below the pipe may create a very undesirable "hard-point effect," as a result of which the structure may suffer distress, especially in the long term. In such cases, the ledge must be removed and replaced with well-compacted granular fill. Alternatively, the rock ledge should be shattered to a minimum depth of about 300 mm (1 ft) below the invert level.

A transition of the pipe foundation from soft ground to rock is also undesirable and should be avoided wherever possible. If it is not feasible to avoid such a transition, the problem can sometimes be corrected by providing a relatively compressible bedding where the foundation is unyielding, and a relatively well-compacted bedding where the foundation is more yielding in nature.

6.2.3   Dewatering

For soil-steel bridges over water courses, the excavation usually extends below the stream bed, and, as a result, the excavated ground can be extremely wet. The water table should, therefore, be lowered before the placing of the bedding material.

The selection of the most appropriate dewatering technique depends upon several factors, such as the size and geometry of the structure, ground profile, type of soil, and the depth of water table. A competent geotechnical investigation is essential, both for making such a selection and also for identifying potential problems that are associated with dewatering. Such problems are discussed briefly in the following paragraphs. It should be noted that for excavations below the water table, computations must be made to determine the rate of drawdown, pumping quantities, and the required pump capacity, as this information is essential for contractors.

In sandy, or similar, cohesionless soils, the construction of a deep sump for dewatering may lead to "boiling" of the soil because of unbalanced hydrostatic heads. When boiling occurs, the sand soil will lose its inherent strength, and this will adversely affect the capacity of the foundation to support the structure. In such soils, the general rule of thumb for the provision of a positive cutoff against seepage into an excavation is to install an interlocking sheet piling, or similar cutoff wall, to a depth which is equal to or more than the unbalanced hydrostatic head above the base of the excavation.

In clayey or cohesive soils, gravity drainage may be totally ineffective because of their low hydraulic conductivity and seepage characteristics. If the construction period is relatively short, dewatering in such soils may not be necessary at all. However, it is very important to bear in mind that the danger of basal heave exists in these soils; such heave may be caused by imbalance in hydrostatic heads and excavation-related stress relief.

In cohesive soils such as silty clays and clayey silt glacial tills, seepage inflow into excavations is minimal and, as a result, pumping out of the water from strategically located sumps may suffice in most cases. It is very important that submersible pumps used for such dewatering should have a minimum capacity of 3 liters/sec (0.26 gal/sec).

In glacial cohesive soils there is always the possibility of the presence of water-bearing silt and sand lenses or pockets. Identification of such geological features is an essential component of the geotechnical investigation.

Once the water table has been lowered, the construction of the soil-steel bridge can proceed on the dry ground, provided that provisions are made to handle the surface runoff. In the case of water course crossings, cofferdams constructed of native soil, or timber or steel sheetings may be useful. It is usual to permit the contractor to select the material for such cofferdams. If cofferdams are required, consideration must be given to maintaining the stream flow during construction. The cofferdam must be designed not only to provide a physical barrier which is stable under the forces of the flowing stream but also to control seepage below its base so as to prevent failure by piping. A basal key is often found useful in providing resistance against lateral instability, as well as for increasing the safety margin against basal seepage or uplift.

The selection of the method of dewatering is generally a contractor's prerogative. However, the method proposed by the contractor must be checked with reference to the factors and potential problems outlined herein before approval is given.

In some cases, where the ground is not very wet, the foundation can be stabilized by placing heavy pit-run material at the bottom of the excavation. In other cases, the heavy pit-run material, used in conjunction with geotextiles or filter cloth, can firm up the foundation adequately.

6.3 Bedding

As defined in Chap. 1, the bedding is the prepared portion of the engineered soil on which the bottom segments of the conduit wall are placed. It can be readily appreciated that the conduit wall should be in direct contact with the soil at all points on its periphery. Such contact is provided for the bottom segments of the pipe by making the bedding of loose granular material and by preshaping the profile of the bedding to the same shape as that of the bottom segments. The upper 150-mm (6 in) layer of the bedding is kept loose in order to ensure that the bottom segments are nested snugly in the bedding, with both the ridges and valleys of the corrugations maintaining contact with the bedding material. Hand-tamping and water-jetting are often required for compacting the soil in hard-to-reach areas.

As shown in Fig. 6.5, the lower portion of the bedding should be lightly compacted to make it relatively firm, and the top 75- to 150-mm- (3 to 6 in) thick layer should be kept quite loose. The compacted portion of the bedding helps to fix the shape of the lower segment of the pipe, and the upper portion ensures a proper contact between the soil and the conduit wall. The bedding material should be free of gravel larger than 75 mm (3 in) in size, and should also be especially free of chunks of highly plastic and organic matter. The curved portion of the bedding should extend laterally to the points at which the radii of curvature of the conduit wall change. In the case of round pipes, for which this criterion cannot be applied for obvious reasons, the bedding should extend to points beyond which there is enough room under the haunches to permit adequate compaction of the backfilling material.

6.3.1 Preshaping

There are two common methods of preshaping the bedding. In one method, which is followed where the curvature of the conduit invert is

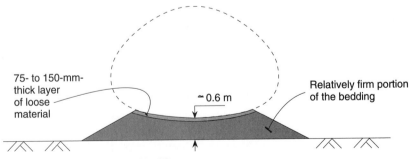

Figure 6.5 Details of preferred bedding.

not very flat, the cross section of the bedding is profiled first to a V-shape. The final profile is obtained usually with the help of a wooden template by adding the bedding material in the middle of the V-shaped profile. This process is illustrated schematically in Fig. 6.6. A photograph of a preshaped bedding can be seen in Chap. 1, Fig. 1.8.

When the invert of the conduit is nearly flat, as it is in the case of pipe-arches, the bedding profile is first made horizontal, as shown in Fig. 6.7. The final profile is then obtained by removing the excess material. Once again, a wooden template can be very helpful in profiling the bedding to the required shape.

6.3.2 Camber

The soil cover above the conduit varies along the pipe (as shown in Fig. 6.8) because of embankment slopes. Due to this uneven cover, the foundation under the pipe settles more under the middle lengths of the conduit than under the outer lengths. If the longitudinal profile of the bedding does not account for this uneven settlement, the conduit invert may develop a sag in the longitudinal direction; this is especially undesirable for conduits conveying water.

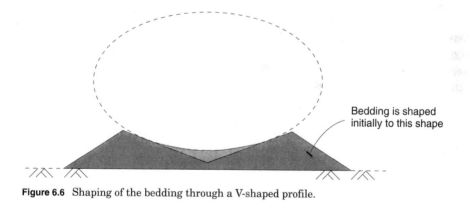

Figure 6.6 Shaping of the bedding through a V-shaped profile.

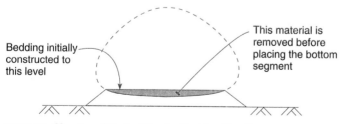

Figure 6.7 Shaping of the bedding by first leveling it.

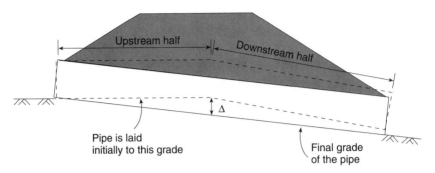

Figure 6.8 Providing initial camber for a structure built over a water course.

To avoid a sag in the longitudinal profile of the invert, the pipe is laid at a camber which compensates for the eventual uneven settlement of the foundation. As illustrated in Fig. 6.8, it is usual to lay the upstream half of the pipe on a flatter grade, and the downstream half on a grade which is steeper than the final required grade of the pipe. If the initial grade of the downstream half of the pipe is selected correctly, the pipe invert will assume a straight longitudinal profile after the foundation has settled.

The corrective cambering of the pipe in the foregoing discussion (illustrated in Fig. 6.8) is clearly for structures built over water courses. In the case of grade separation structures, which may require a nearly horizontal invert, the precambering may have to be done as illustrated schematically in Fig. 6.9.

The camber is quantified by the distance Δ, which is illustrated in Figs. 6.8 and 6.9. A reliable estimate of the required value of Δ can be obtained only by an analysis of the foundation by taking account of the geometry of the conduit, the compressibility of the foundation soils, and the depth of soil cover over the conduit. However, it may be useful to note that the soil-steel bridge industry generally uses a rule of thumb according to which Δ is taken as a minimum of 0.5 percent of the length of the pipe, irrespective of the type of foundation and the depth of cover.

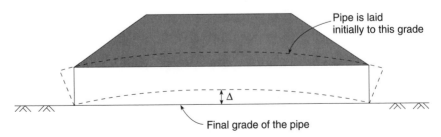

Figure 6.9 Providing initial camber for a grade separation structure.

6.4 Construction of the Pipe

As discussed in Chap. 1, this book relates only to those structures in which the metallic shell or pipe is assembled on the site by bolting together curved corrugated steel plates with annular corrugations which generally have a profile of 152×51 mm (6×2 in). The pipes of these structures are generally constructed either by a plate-by-plate assembly method, or by a component subassembly method. These two methods are described in this section. There are other methods in which partial or full lengths of the pipe are assembled away from the bedding; however such methods are not described here.

6.4.1 Lapping pattern

The longitudinal and circumferential seams of the pipe are illustrated in Chap. 1, Fig. 1.15. The lapping pattern at the various seams has an influence not only on the ease of construction, but also on the structural integrity. Experience has shown that: (1) a pipe with a few continuous longitudinal seams is easier to construct than one with all longitudinal seams staggered; and (2) it is undesirable to have four plates meeting at a corner. A pipe with staggered longitudinal seams is preferred because it is not prone to "zipper" types of failures that have been associated with pipes having continuous longitudinal seams.

6.4.2 Selection of plate lengths

Corrugated steel plates of 152×51 mm (6×2 in) corrugation profile come in two standard net lengths; 3.05 m (10 ft) and 3.66 m (12 ft). It is recalled that, as shown in Fig. 1.17, the net length of a corrugated plate is the distance between the centerlines of the two rows of bolt holes for circumferential seams. Since pipes are constructed from standard plate lengths of 3.05 m (10 ft) and 3.66 m (12 ft), the total lengths of 6.10, 7.32, 9.14, 9.75, 10.36, 10.97, and 12.19 m (20, 24, 30, 32, 34, 36, and 40 ft, respectively) can be assembled readily. Larger lengths are in multiples of 0.61 m (2 ft). It should be noted that nonstandard lengths can indeed be used. However, this practice leads to extra cutting and wastage.

If all the plates for a pipe have the same length and some continuous longitudinal seams, then, as shown in Fig. 6.10a, it is difficult to avoid the undesirable situation of having four plates meet at a corner. For this reason, it is advisable to select the length of the pipe in such a way that for each section there is at least one plate whose length is different from the lengths of the other plates. For example, a pipe with a length of 12.80 m (42 ft) is preferable to one with a length of 12.20 m (40 ft) simply because the former pipe can have three plates of length

3.05 m (10 ft), and one of a length of 3.66 m (12 ft), while the latter pipe can be constructed only with plates of the same length; these two cases are illustrated in Fig. 6.10b and a, respectively. In some cases, the length and slope of the beveled section of a barrel can be selected to avoid the four corner laps.

6.4.3 Selection of plate widths

It is recalled that the plate width is the dimension measured along the circumference of the conduit and that the net width of the plate is measured between the centerlines of the two rows of bolts for the longitudinal seams. As discussed in Chap. 1 and shown in Table 1.1, the plates with 152 × 51 mm (6 × 2 in) corrugation profile are available readily in only three standard widths; these are designated as 3N, 6N and 9N plates, whose net widths are 732, 1463, and 2195 mm (28.8, 56.6, and 86.4 in), respectively. These plates have 6, 7, and 10 holes, respectively, along the curved portion of the plate, and are also designated as 6H, 7H, and 10H plates, respectively. Holes along the curved portion of the plate are spaced at 244 mm (9.61 in) c/c measured along the circumference.

The following guidelines will be found helpful for selecting plate widths:

- Minimize the number of plates to construct the ring.
- Ensure that the longitudinal seams are staggered within a segment of constant radius.
- Ensure that the radius of curvature of each segment is consistent with the radius used in design.

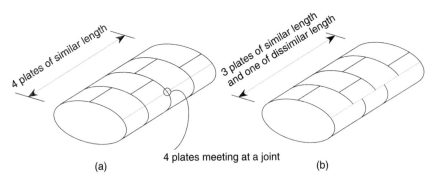

Figure 6.10 Two examples of pipe length selection: (a) undesirable arrangement; (b) desirable arrangement.

The plate suppliers tend to rotate their stock, because of which it is desirable to be flexible enough to accommodate alternative plate combinations.

Two examples of plate-width selection for the top segment of a horizontally elliptical pipe are shown in Fig. 6.11a, b. The former figure shows a desirable arrangement in which the longitudinal seams within the segment are staggered by selecting plates of different widths for successive segments in the longitudinal direction; the latter figure shows an undesirable arrangement in which the longitudinal seams within the segment are continuous because of selecting the same plate widths for the entire length of the pipe.

For calculating the cross-sectional area and the circumference of horizontally and vertically elliptical conduits, it is assumed that the conduit shape is truly elliptical, although, as explained in Chap. 1, strictly speaking it is not. The error involved in making this assumption is very small. The cross-sectional area A and circumference C of an ellipse are given by

$$A = 0.25\pi\, D_h D_v \tag{6.1}$$

$$C \simeq 0.50\pi(D_h + D_v)\, \frac{64 - 3m^4}{64 - 16m^2} \tag{6.2}$$

where D_h and D_v are the span and rise, respectively, as shown in Fig. 6.12, and m defines the conduit shape by

$$m = \frac{D_h - D_v}{D_h + D_v} \tag{6.3}$$

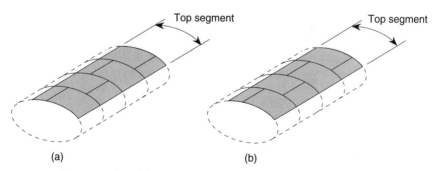

Figure 6.11 Two examples of plate width selection for the top segment of a pipe: (*a*) desirable selection with staggered longitudinal seams; (*b*) undesirable selection with continuous longitudinal seams.

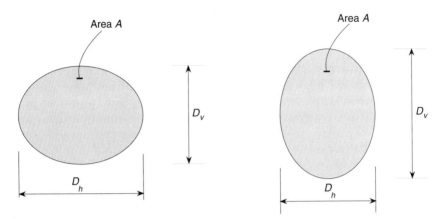

Figure 6.12 Notation for span and rise.

Having established the shape of the conduit and its rise and span, the circumference of the conduit wall, C, can be determined approximately from Eq. (6.2) or other similar equation when the conduit shape is other than circular or elliptical.

By converting the circumference of the conduit to inch units and dividing it by 9.6, the circumference is obtained in measures of N which is familiar to the North American corrugated plate industry and which has been explained in Chap. 1, Sec. 1.2.

The process of selecting plate widths can be explained with the help of the example of a horizontally elliptical pipe whose span D_h and rise D_v are determined to be 7.01 and 4.29 m, respectively. From Eq. (6.2), the circumference C of this conduit is found to be 18.00 m or 709 in. Dividing 709 by 9.6, one gets 73.85, which is rounded to 74. Hence, in terms of the industry notation, the required circumference is 74N. The horizontally elliptical pipe is composed of four segments, namely, two side segments, and the top and bottom segments. These segments can have a variety of circumferential lengths. For example, each of the top and bottom segments can have a circumferential length of 27N, and each of the side segments 10N. In this case, the top and bottom segments can be composed of three 6N plates and one 9N plate so as to ensure that the longitudinal seams within the segment are staggered, as shown in Fig. 6.13. The side segments, because of having to be formed out of two 5N plates, cannot have staggered longitudinal seams.

In such a case, it is desirable to change the circumferential length of the side segments even if it leads to a slightly smaller or larger pipe. For example, for the case under consideration, the longitudinal seams in the side segments can be avoided completely by using a 9N plate

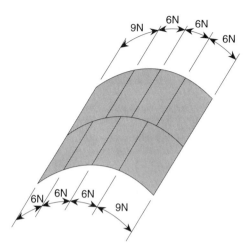

Figure 6.13 Example of selection of plate widths for the top segment of a horizontally elliptical pipe.

and, consequently, reducing the circumference slightly; alternatively, the circumference can be increased slightly by using one 6N and one 5N plate, in which case the longitudinal seams in the side segments can also be staggered.

6.4.4   Bolting arrangement

Steel plates with 152×51 mm (6×2 in) corrugations are bolted together through bolts of 20-mm (¾ in) nominal diameter. The bolts and nuts are made of heat-treated steel meeting the requirements of ASTM Standard F568, and are galvanized. The insides of the bolt heads and one side of the nuts are chamfered. The bolt holes in the steel plates have a nominal diameter of 25 mm (1 in).

As shown in Chap. 1, Fig. 1.18, all longitudinal seams have bolts in two rows which are spaced at 51 mm (2 in) center to center. An extensive discussion on the strengths of various bolting arrangements is given in App. B. However, it may be noted here that the preferable bolting arrangement is the one shown in Fig. 1.18a (Chap. 1) and that in this arrangement the bolts in the row closer to a visible edge of the lapping plates are placed in the valleys, as shown in Fig. 6.14.

To permit retorquing of the bolts after the backfill has been placed around the pipe, the nuts are sometimes kept toward the inside of the conduit. However, in certain applications, such as pedestrian or cattle underpasses, the bolt heads are kept toward the inside of the conduit because of safety and aesthetic considerations.

Bolts are usually supplied in two lengths, with the shorter bolts being for connecting two plates at their seams, and the longer ones for three plates. The bolts are long enough to allow full threading at the

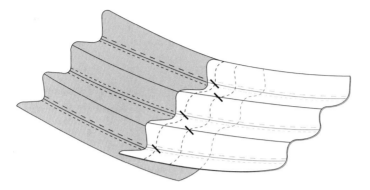

Figure 6.14 A desirable bolting arrangement.

nominal torque of about 200 N · m (147.5 lb · ft) and the maximum torque of about 400 N · m (295 lb · ft). At the final torque, two or more threads of the bolt shaft are visible beyond the nut. It is important that the nuts be oriented so that their chamfered sides are in contact with the plates.

6.4.5 Nesting of seams

If the lapping plates at a longitudinal seam are not nested properly, excessive bending may take place during the assembly of the pipe. The possibility of having improperly nested plates is particularly great for longitudinal seams at the haunches of pipe-arches, where the mating plates have markedly different radii of curvature. Figure 6.15*a* shows an improperly nested seam in which the mating plates are clearly incompatible at the seam. A desirable arrangement of the plate curvatures at the seam is one in which the lapping portions of the adjacent plates are straight and at compatible inclinations, as shown in Fig. 6.15*b*.

6.4.6 Plate-by-plate assembly

A common method of constructing the metallic shell or pipe of a soil-steel bridge is the plate-by-plate assembly method. In this method the bottom segment of the pipe is first assembled on the shaped bedding as described earlier. This first sequence of assembly is illustrated schematically in Fig. 6.16*a*. It is noted that it is usually easier to pre-assemble the bottom segment and then place it on the shaped bedding.

In the next sequence, the complete ring of a small length is assembled at one end by first adding the 3.65-m- (12 ft) long side-segment plates to the bottom segment and then completing the ring by adding

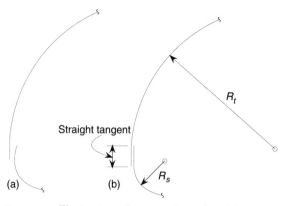

Figure 6.15 Illustrations of seams at haunches: (*a*) improperly nested seam; (*b*) properly nested seam.

the 3.05-m (10 ft) top-segment plates. This ring is bolted loosely and adjusted to shape. The complete ring at one end provides some stability and reference for further assembly. The longer side plates facilitate the closure of the ring. As shown in Fig. 6.16*b,* when the ring at one end is completed, further side plates are added and subsequent rings closed in sequence.

Figure 6.16*c* shows further sequential progress in the assembly of the pipe. It can be seen in this figure that different lengths of the pipe are at different stages of completion. It is not a recommended practice to connect all the side-segment plates along the pipe length in one operation. The preferred practice is as illustrated in Fig. 6.16*c.* The progressive building up of the pipe continues until the whole pipe is assembled, as shown in Fig. 6.16*d.*

Figure 6.16 also shows the preferred practice of adding plates symmetrically to both sides of the bottom segment. If this practice is not followed (i.e., if plates are added to only one side), the pipe is likely to rotate on the bedding.

To start with, plates should be connected by using three or four bolts for each longitudinal seam. These bolts, which are located near the middle of each seam, should be lightly tightened initially so that the pipe has enough flexibility to facilitate the joining of other plates.

After a part of the pipe has been assembled, the remainder of the bolts can be inserted and tightened lightly. It is advisable to work from the center of the plates toward the corner, with the bolts in the corner holes being inserted last. The purpose of this procedure is to encourage the gradual nesting of all the seams so that a large gap does not occur in the final tightened seam. The tightening of all the bolts should commence systematically from one end of the pipe to the other after all the plates are assembled, all the laps are nested, and all the bolts are in

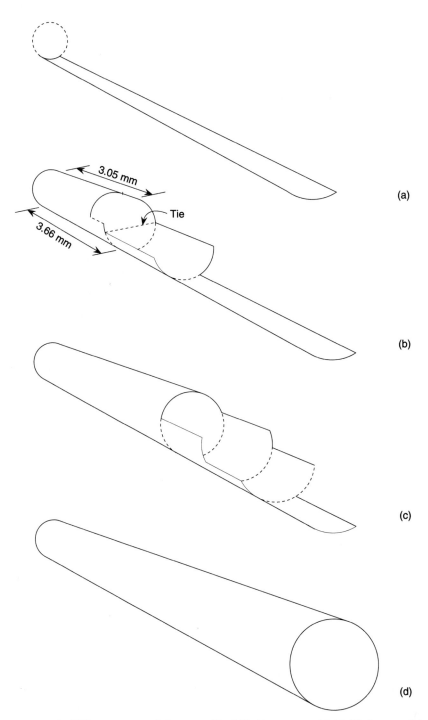

3.05 mm

Tie

3.66 mm

(a)

(b)

(c)

(d)

Figure 6.16 Different stages of pipe assembly: (*a*) bottom segment laid; (*b*) ring completed for partial length; (*c*) ring completed for partial length and side walls erected for some of the remaining length; (*d*) pipe completed.

place. Tightening of all the bolts can result in the loosening of previously tightened bolts. Therefore, it is advisable to repeat the entire sequence of tightening the bolts, after which the bolt torque should be checked. For best results, 5 percent of all the bolts should be checked randomly to ensure that the torque in at least 90 percent of the tested bolts is within the range of 200 to 400 N · m (147.5 to 295 lb · ft).

Care should be taken to avoid overtorquing of the bolts, which can create stress-raisers in the plates by denting them. The use of correct tools is helpful in avoiding such overtorquing. A small impact wrench with a 12-mm (½ in) drive is often quite suitable for initial torquing, while the final torquing can be done with a wrench of 18-mm (¾ in) drive. An impact wrench of 25-mm (1 in) drive can easily overtorque the bolts and should therefore be used with great caution.

It is important that the required cross-sectional shape of the pipe be maintained at all stages of assembly. This can be achieved conveniently with the help of sizing cables which control the chord lengths of the conduit, and sizing props which maintain the conduit rise. As shown in Fig. 6.17a, in exceptional circumstances the shape of the bottom segment can be maintained by controlling its chord length through sizing cables. When the side-segment plates are added to the bottom segment, the sizing cables are installed as shown in Fig. 6.17b. If the shape is adjusted properly during assembly the top segment should fit in snugly, without the need for forcing the side segments toward each other.

For larger-span structures, the pipe can deform considerably under its own weight. In this case, sizing props are used to maintain the rise, as shown in Fig. 6.17c. It is important to ensure that the cables and props do not hinder the deformation of the pipe during the backfilling operation. Accordingly, the props should not be mechanically connected to the crown; this will permit the upper portion of the pipe to move up freely as the backfill is added to both sides of the conduit. The props must be self-supporting and stable so that they do not fall when the crown moves up. As the backfill is added above the level of the con-

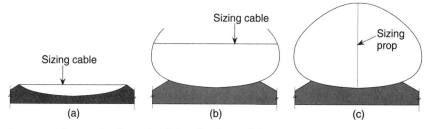

Figure 6.17 Accessories for maintaining the shape of the pipe during assembly: (a) sizing cable for bottom segment; (b) sizing cable for side segments; (c) sizing prop for completed pipe.

duit shoulders, the side segments of the conduit wall are forced out-
ward toward the soil; when this stage is reached in the backfilling
operation, the sizing cables should be removed. Enough props and
cables should be provided to avoid local damage in the vicinity of these
temporary supports.

Before starting the backfilling operation, checks should be made to
ensure that the cross-sectional shape of the pipe is not significantly different from the design shape. Loosening of bolts, readjusting of the
shape, and other corrective measures should be taken if the shape is
unacceptable.

6.4.7 Component subassembly

The method of component subassembly for constructing the pipe is different from the plate-by-plate assembly method only in that it permits
the assembly of some pipe segments away from the bedding. This
method is particularly useful for reducing the construction time for
large structures. The pipe segments are assembled in a less restrictive
area, and advantage can be taken of gravity in nesting the seams properly. Further, efficiency of the method lies in permitting several different operations at the same time. For example, the bedding can be
prepared at the same time that the bottom segments of the pipe are
assembled in the subassembly area, shown in Figs. 6.2 and 6.3. The various segments of the pipe which are subassembled away from the pipe
location should not be too large for the crane being used. Other principles and procedures for assembling of a pipe by the component subassembly method are the same as those for the plate-by-plate assembly
method which was described earlier. Since poorly nested laps are common when this method is used, careful inspection is recommended.

6.5 Backfilling

It has been noted previously that the backfill is a limited portion of the
soil around the conduit, excluding the bedding. The backfill has a very
significant influence on the structural integrity of a soil-steel bridge
and, accordingly, it is important that it be composed of appropriate
material and that it should be compacted adequately. Chapter 4 has
dealt with the selection of the backfill materials and the quantification
of compaction. The procedures for placing and compacting the backfill
are discussed following.

6.5.1 Extent of backfill

The high-quality granular material required for the backfill is usually
more expensive than material of lower quality, but the latter may not
be suitable for use in the immediate vicinity of the pipe. To achieve

economy, it is therefore necessary to limit the extent of the high-quality backfill to the zones immediately adjacent to the structure and beyond which a lower-quality material will suffice. Following good construction practice, and also as required by some codes (e.g., the Ontario Highway Bridge Design Code), the backfill should extend transversely to at least half the span on each side of the conduit, and vertically to the minimum required depth of cover above the crown. The minimum extent of backfill thus defined is shown in Fig. 6.18; this should be shown on construction drawings.

It is recalled that, as discussed in Chap. 5, the Ontario code requires the minimum depth of cover to be larger of $(D_h/6)$ $(D_h/D_v)^{0.5}$ and $0.4(D_h/D_v)^2$, where D_h and D_v are the span and rise, respectively, in meters, it being noted that for pipe-arches and arch structures, D_v is defined as illustrated in Fig. 6.19. The depth of cover should not in any case be less than 0.6 m.

Beyond the high-quality backfill zones shown in Fig. 6.18, lesser-quality material can be used; however, it is usually not feasible to limit the good-quality backfill only to the zones shown. For practical reasons, the good-quality backfill material is provided even beyond the specified zones, especially in the transverse direction. One example of the optimum use of the good-quality and lower-quality material can be seen in Chap. 1, Fig. 1.24.

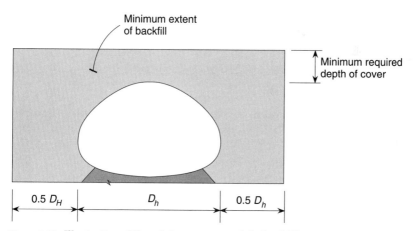

Figure 6.18 Illustration of the minimum extent of the backfill.

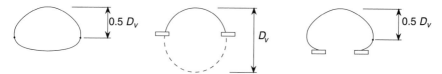

Figure 6.19 Definition of D_v for the calculation of the minimum depth of cover.

6.5.2 Backfill under haunches

The backfill under the haunches should be placed in layers not exceeding 150 mm (6 in) in thickness when compacted. The soil under the
haunches of some structures, in particular pipe-arches, has to be compacted more than the rest of the backfill. Accordingly, it is important
that particular care should be exercised in compacting this portion of
the backfill. Since the area under the haunches is difficult to reach by
the conventional compacting equipment, the soil in that region should
be compacted manually by means of timber rammers and hand-held,
air-driven compactors; this will ensure that the soil is compacted adequately, especially between the corrugations and other hard-to-reach
areas under the haunches.

For compaction under the haunches, the hand-held compactors
should weigh not less than 9 kg (20 lb), and should have a tamping face
not larger than 150 × 150 mm (6 × 6 in). Ordinary sidewalk compactors
are usually too light to be effective for this compaction.

Water-jetting is a specialized technique for compacting the soil
under the haunches. However, faulty use of this technique can lead to
problems, and hence its use is recommended only under the direct
supervision of an experienced inspector. Controlled low-strength material (CLSM) discussed in Chap. 8 can be used effectively, especially
under the haunches, to provide the soil support that is adequately stiff.

As for the remaining backfill, the soil should be added and compacted under the haunches in a symmetrical fashion on both sides of
the conduit. Failure to do so will result in rotation of the pipe, usually
referred to as *rolling*.

6.5.3 Placement of backfill

The backfill material should be placed and compacted in layers not
exceeding 300 mm (12 in) of compacted thickness, with each layer
being compacted to the required density at the optimum moisture content. Layers thinner than 300 mm (12 in) will normally lead to even
more uniformly compacted backfill. It is noted that some jurisdictions
require a thickness of 150 mm (6 in) for compacted layers. The layers
of backfill should be placed and compacted on both sides of the conduit
in such a way that, at any time, the difference in the levels of the backfill on the two sides at a given transverse section does not exceed twice
the thickness of the typical compacted layer.

Ideally, the backfill should rise evenly on both sides of the conduit.
However, the foregoing recommendations are considered to be a reasonable compromise for optimizing the efficiency of the construction.

Stones exceeding 80 mm (3 in) in size should not be permitted within
300 mm (12 in) of the conduit wall. In order to limit the development of

incremental pressures that may cause excessive deformation of the conduit wall during the backfilling operation, only smaller compaction equipment should be allowed to operate very close to the pipe, say, within 600 mm (24 in) of it.

The equipment delivering the backfill soil should not be permitted to travel in the transverse direction in the vicinity of the pipe, because the dumping of the material in this manner makes spreading of the soil awkward and can lead to pockets of uncompacted backfill adjacent to the pipe. The undesirable positioning of the trucks delivering backfill material near the pipe is illustrated in Fig. 6.20. To ensure that uncompacted pockets of soil are not left in the vicinity of the pipe, trucks delivering the backfill material should always travel parallel to the axis of the pipe as shown in Fig. 6.21. Similarly, it is preferable for the compacting equipment to travel along the pipe.

The placing and compacting of the backfill around the pipe can be achieved effectively and efficiently by following and repeating this sequence:

1. Use dump trucks or scrapers to deliver sufficient backfilling soil for one layer, on each side of the pipe.

2. Use graders or dozers for spreading evenly the layer of soil.

3. Use light, hand-held compactors to compact the layer of soil in the immediate vicinity of the pipe, and heavier self-propelled vibratory drum compactors for the rest of the area.

It is important to prepare a sieve analysis and conduct a Proctor test prior to the delivery of the backfill. On-site testing should preferably

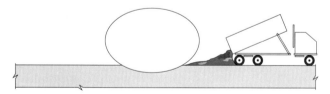

Figure 6.20 Undesirable manner of delivering backfill material.

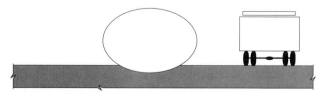

Figure 6.21 Desirable traveling position of equipment.

include a nuclear density gauge to establish the rolling pattern and to confirm moisture content and compaction achieved.

6.5.4 Placing backfill over the top

After the fill has reached the top of the side segments of the pipe, only light equipment should be used to spread and compact the backfill over the top portion of the pipe. In some cases even manual placement may be preferable. As can be appreciated, the top segment of the conduit wall, without substantial compacted soil around it, is very susceptible to damage by the construction equipment. The fill over the top portions of the pipe should be placed and compacted in the transverse direction of the pipe. Initial layers of soil should be placed loose to the shape of the top segment as shown in Fig. 6.22. Normal backfilling procedures can be used after the fill above the crown has reached the level corresponding to the minimum depth of cover requirements which are discussed earlier in the chapter.

6.5.5 Control of pipe deformation

As was noted in Chap. 1, the metallic shell of a soil-steel bridge without the backfill all around it is very flexible. Hence, it is susceptible to significant deformations during the backfilling operation. Moderate deformations of the pipe are indeed necessary for the composite action to develop between the conduit wall and soil; however, large deformations which can cause permanent set in the conduit wall should be avoided.

There are three main types of deformation that are of concern in the pipe. One type is often referred to as the *rolling deformation,* in which the pipe rotates about its axis as illustrated in Fig. 6.23a. This type of deformation is caused by the backfill pressure on one side of the pipe being larger than on the other side. To avoid rolling deformation, the difference between the levels of the backfill on both sides of the pipe should be kept to a practical minimum.

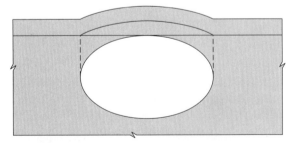

Figure 6.22 Initial layers of backfill over the pipe.

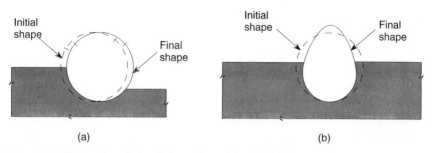

Figure 6.23 Illustration of pipe deformations: (*a*) rolling deformation; (*b*) peaking deformation.

The second type of deformation is referred to as the *peaking deformation,* and is illustrated in Fig. 6.23*b*. This deformation is caused by the horizontal pressures exerted by the backfill before it reaches the shoulder level of the conduit. The various methods of predicting the peaking movement of the crown during backfilling were discussed in Chap. 5.

The peaking of the crown can be controlled either by reducing the compaction effort in the vicinity of the pipe or by placing soil above the pipe. The latter procedure, which can be seen in operation in the photograph of Fig. 6.24, is practiced quite commonly. In exceptional circumstances (e.g., in the case of high-profile arches), the peaking can also be controlled by piling earth against the conduit wall inside the conduit, as shown schematically in Fig. 6.25. The soil inside the conduit is, indeed, removed after the pipe shape has been stabilized with the backfill all around the conduit.

The third type of deformation relates to the *failure of the soil cover* as discussed in Chap. 4; it is caused by eccentric heavy loading over the pipe before the depth of cover has reached the required minimum depth. This kind of deformation can affect structural integrity and should be avoided at all costs. Careless use of the construction equipment can also cause localized deformations such as dents, bulges, tears, and scrapes. These deformations, although not always critical, should also be avoided.

6.5.6   Monitoring deformation

The deformation of the pipe during the backfilling operation can be monitored conveniently and effectively by means of three plumb bobs attached (according to the general scheme of Fig. 6.26) at every transverse section of the pipe at which deformations are to be checked. In most cases it is sufficient to monitor the pipe shape at three transverse sections; one being at midlength and one near each end of the pipe. The plumb bobs are so hung that they are a specific distance from corre-

Figure 6.24 Surcharging over the pipe to control peaking.

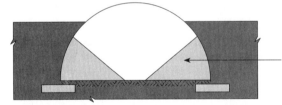

Temporary soil support
for the conduit wall

Figure 6.25 A method of controlling peaking deformation.

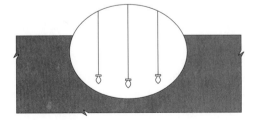

Figure 6.26 Monitoring pipe deformations through plumb bobs.

sponding points marked at the bottom of the pipe. The peaking deformation can be detected by the upward movement of the plumb bobs, and the rolling deformation by their lateral movement.

When the conduit carries traffic even before the placing of the backfill around it (as may be the case, for example, with railway underpasses), it may not be convenient to let the plumb bobs hang inside the conduit for extended periods. In this case, targets should be installed inside the pipe at the specified sections, and their movement monitored by means of theodolites placed outside the pipe.

6.5.7   Site supervision

Close and competent site supervision is essential to ensure that a soil-steel bridge is constructed properly. Approval of the backfill material, concrete mix design, scaffolding design and other construction items should be completed before the construction begins. For larger structures, namely, those with spans larger than about 8 m (26 ft), continuous inspection is recommended at all stages of construction. For smaller structures, the construction work should be inspected at each of the following stages of construction:

1. at the completion of the excavation of the foundation

2. at the completion of the shaping of the bedding

3. at the completion of the erection of the pipe

4. during the compaction of the backfill under the haunches

5. after the backfill has reached the spring-line level

6. after the backfill has reached the crown level

7. during the placing and compaction of the backfill over the top portion of the conduit until the fill has reached the level corresponding to the minimum depth of cover

The authorization to continue work for the next stage should be given only after the work for the previous stage has been carried out according to the specifications. Of course, the quality and compaction of the soil used in the backfill should be inspected carefully to ensure that it meets the specifications.

7

Modeling of Soil

George Abdel-Sayed
and Bozena B. Budkowska
University of Windsor
Windsor, Ontario, Canada

The purpose of this chapter is to: (1) acquaint the reader with the essential terminology and concepts related to the backfill of a soil-steel bridge; (2) discuss the fundamentals of the theory of elasticity as applied to the analysis of soils; and (3) provide guidance for obtaining the values of the relevant parameters used in the design and analysis of these structures by simplified methods.

7.1 Introductory Concepts

As is the case in most structures, the mathematical modeling of the different components of a soil-steel bridge is the first essential step in the design process and the formulation of acceptable criteria for the load-carrying capacity of the completed structure.

Many investigators have attempted to model the effect of the soil component, employing various analytical and empirical techniques. The complexities of these models depend upon the rigor of the methods of analysis; these may include finite element (FE) methods and approximate, simplified methods. The former methods can incorporate non-linear, elastic-plastic, and stress-dependent properties of the soil, while the latter typically express the soil properties in terms of a limited number of parameters.

The soil in a soil-steel bridge constitutes a structural component; however, it differs from conventional structural materials in several ways. For example, the conduit wall is made of a material whose chemical and physical properties are closely controlled and defined. By contrast, the soil (being a natural material) has physical properties which vary very widely from one location to another. Since very large quantities of soil are needed in a soil-steel bridge, it is not always feasible to transport the good, uniform-quality material that is preferred over more than relatively short distances. Nor is it practical to process all of the soil. The result of these difficulties is that the soil component of the composite bridge can vary in properties from site to site, depending upon its source.

Preferably the designer of a soil-steel bridge should be able to study the soil at nearby sources of the backfill material, and identify the required effort to improve its behavior when placed in the structure. It is essential that the designer be able to evaluate those properties of the soils which are needed for the design and analysis of the structure.

7.1.1 Classification of soil particles

The sizes of the soil particles and their distribution are the main factors which influence the mechanical properties of the soil. For convenience in expressing the size characteristics of the soil components, a generally acceptable notation has been developed for defining particles of different size ranges. The ASTM notation (Ref. 3), which is used by many agencies, is defined in Table 7.1, it being noted that some agencies have slightly different size ranges for defining soil particles of a certain group.

7.1.2 Grading of soil

The distribution of different particle sizes in a soil mass is referred to as the *soil grading*. For the sand and gravel portion of the soil, the grading is established by passing it through sieves of different sizes. For clayey and smaller-particle-size portions, the different particle sizes are separated by measuring the settling velocities of the particles when the soil sample is allowed to settle in a water suspension.

The grading of the soil is often expressed as the percentages of the total dry weight of the particles which can pass through sieves of different sizes. This soil grading can be presented graphically by means of the maximum particle-size distribution curve, two examples of which are shown in Fig. 7.1. As can be seen in this figure, the maximum par-

TABLE 7.1   Size Classification of Soil Particles

	Particle Size Limits in Millimeters for:					
Agency	Gravel	Coarse sand	Fine sand	Silt	Clay	Colloidal
AASHTO	76.2-2.0	2.0-0.42	0.42-0.075	0.075-0.002	<0.002	<0.001
ASTM	76.2-2.0	2.0-0.42	0.42-0.074	0.074-0.005	<0.005	<0.001

ticle size is drawn on a logarithmic scale. The shape of a grading curve defines the soil group; for example, it can be appreciated that a smooth and well-spaced-out curve represents a well-graded granular material, while, on the other hand, a curve limited to the left-hand side of the particle size scale represents a predominantly clayey soil.

The grain-size distribution curves are very useful in comparative analysis of various types of soils. On the basis of the soil gradation curves two basic parameters can be determined, i.e., uniformity coefficient and gradation coefficient. These parameters are used in classification of soils—as, for example, in the Unified Soil Classification System (USCS). The *uniformity coefficient, C_u,* is defined as:

$$C_u = \frac{D_{60}}{D_{10}} \qquad (7.1)$$

where D_{10} is the diameter corresponding to 10 percent fines in the grain-size distribution curve and D_{60} is the diameter corresponding to 60 percent fines in the grain-size distribution curve.

The second parameter, C_c, which is called the *coefficient of gradation* (or *coefficient of concavity*), is defined by:

$$C_c = \frac{D_{30}^2}{D_{60}\,D_{10}} \qquad (7.2)$$

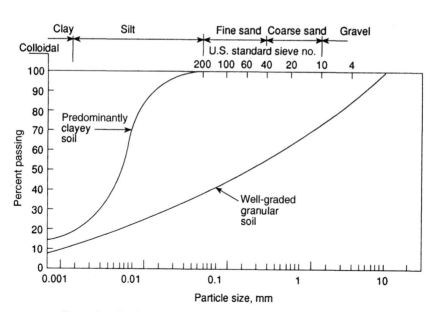

Figure 7.1 Examples of soil grading.

where D_{30} is the diameter corresponding to 30 percent fines in the grain-size distribution curve.

Different jurisdictions have their own designations for different soil gradings. The Ministry of Transportation of Ontario uses the designation of granular A, B, and C soils. Granular A soils are very well graded gravel or sandy gravel soils; granular B soils are also gravel or sandy gravel soils, but may not be as well graded as granular A soils; and granular C soils are clayey gravel or clayey, sandy gravel soils. Granular A and B soils are considered to be suitable for the backfill of a soil-steel bridge, whereas granular C soils are not.

The grading envelopes of granular A, group I soils, and granular B, group I soils (Table 5.2) are shown in Fig. 7.2*a* and *b*, respectively. It can be seen that the former envelope, being relatively narrow, can include only the grading curves of well-graded soils. The latter curve, on the other hand, being much wider, can include the grading curves of even poorly graded soils.

7.1.3 Classification of soil

Many important physical parameters of the soils are associated with soils classification according to Unified Soil Classification System (USCS). For instance, the Ontario Highway Bridge Design Code (OHBDC) (see Table 5.1) employs groups of soils defined by USCS (ASTM designation D-2847).

This soil classification system is based on the grain-size distribution curve and the Casagrande plasticity chart (Fig. 7.3).

The Casagrande plasticity chart is presented in a coordinate system of plasticity index against liquid limit. Employing this chart, the soils are classified based on their plastic properties, which are defined by Atterberg's consistency limits (Ref. 6).

According to the Unified Soil Classification System, the soils are divided into two main groups, i.e., coarse-grained and fine-grained soils.

Coarse-grained soils are defined as those which have 50 percent or more granular material retained on a no. 200 sieve (with opening equal to 0.075 mm). Fine-grained soils are those for which more than 50 percent of the material passes a no. 200 sieve.

The coarse-grained soils are divided into two subgroups:

1. gravels and gravelly soils

2. sands and sandy soils

The former subgroup has the letter G as prefix in group symbols, while the latter has the letter S as prefix.

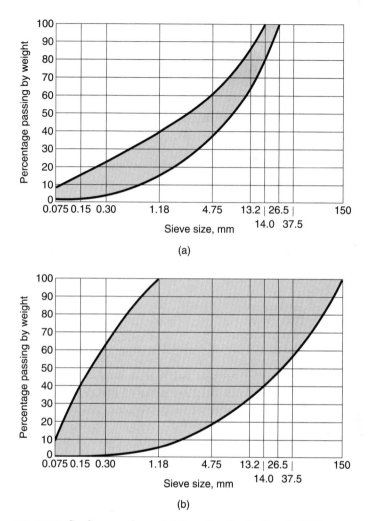

Figure 7.2 Grading envelopes: (*a*) Granular A soils, type I; (*b*) Granular B soils, type I.

The gravels are defined as those which have the greater percentage of coarse-grained fraction retained on a no. 4 sieve (with opening equal to 4.75 mm). The sands are those which have the greater percentage of coarse-grained fraction passing a no. 4 sieve.

Depending on the grain-size distribution curves (regarding coefficients of uniformity C_u and gradation C_c) and the nature of fines (taking into account plastic properties which are analyzed by means of Casagrande plasticity chart), the gravels and sands are subdivided into eight groups:

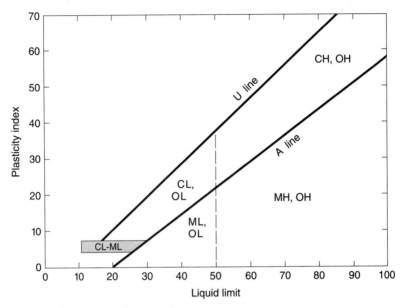

Figure 7.3 Casagrande plasticity chart.

1. Coarse-grained soils (less than 5 percent passing a no. 200 sieve) with $C_u \geq 4$ and $C_c = 1$ to 3 for gravels and $C_u \geq 6$ as well as $C_c = 1$ to 3 for sands are called *well-graded coarse-grained soils* (suffix W in group symbol), and are designated by group symbols GW and SW for gravels and sands, respectively.

2. Coarse-grained soils (less than 5 percent passing a no. 200 sieve) for which the criteria for well-graded soils are not satisfied are called *poorly graded gravels and sands* (suffix P in group symbol), and are denoted by group symbols GP and SP for gravels and sands, respectively.

3. Coarse-grained soils (more than 12 percent passing a no. 200 sieve) with plasticity index PI less than 4 (indicating silt binder) and plastic properties located below the A line in the Casagrande plasticity chart are called *silty gravels and silty sands* (suffix M in group symbol), and are denoted by group symbols GM and SM for gravels and sands respectively.

4. Coarse-grained soils (more than 12 percent passing a no. 200 sieve) with plasticity index PI greater than 7 (indicating clay binders) and with plastic properties placed above A line in the Casagrande plasticity chart are called *clayey gravels and clayey sands* (suffix C in group symbol), and are designated by group symbols GC and SC for gravels and sands, respectively.

For coarse-grained soils (5 to 12 percent passing a no. 200 sieve) a dual notation for group symbols is used. The first part of the group symbol is connected with well- or poorly graded gravels or sands (following the criteria associated with subgroups GW, SW, GP, SP embodied in suffix W or P), while the second part of the group symbol is connected with the nature of the fines and is defined according to the criteria characteristic for silty and clayey gravels and sands regarding plastic properties (i.e., according to the criteria of GM, SM, GC, SC connected with suffix M or C).

Thus the following additional four groups of coarse-grained soils are introduced in USCS which are designated by means of dual-group-symbol notation.

5. Coarse-grained soils which satisfy the criteria for well-graded soils (C_u, C_c) of group (1) with plastic requirements of group (3) (silty soil, plastic properties below A line in the Casagrande plasticity chart) are called *well-graded silty gravels* or *well-graded silty sands,* and are denoted by means of group symbols GW-GM and SW-SM, respectively.

6. Coarse-grained soils satisfying the criteria of poorly graded soils (group 2) with plastic characteristics of group (3) are called *poorly graded silty gravels or poorly graded silty sands,* and are designated by means of group symbols SP-GM and SP-SM, respectively.

7. Coarse-grained soils which satisfy the criteria for well-graded soils of group (1) with plastic characteristics for group (4) (clayey soil, plastic properties above A line in the Casagrande plasticity chart) are called *well-graded clayey gravels* or *well-graded clayey sands,* and are designated by group symbols GW-GC and SW-SC, respectively.

8. Coarse-grained soils satisfying the criteria for poorly graded soils of group (2) with plastic properties of group (4) are called *poorly graded clayey gravels* or *poorly graded clayey sands,* and are denoted by group symbols GP-GC and SP-SC, respectively.

The preceding classification is summarized in Tables 7.2 and 7.3 for determining the group symbol for gravelly and sandy soils, respectively.

The fine-grained soils (more than 50 percent passing a no. 200 sieve) are divided into the following groups:

1. inorganic silts with prefix M in group symbol

2. inorganic clays with prefix C in group symbol

3. organic silts and clays with prefix O in group symbol

TABLE 7.2   Unified Classification System—Group Symbols for Gravelly Soil
(Reproduced with permission from Ref. 6.)

Group symbol	Criteria
GW	Less than 5% passing no. 200 sieve; $C_u = D_{60}/D_{10}$ greater than or equal to 4; and $C_c = (D_{30})^2/D_{10} \times D_{60}$ between 1 and 3
GP	Less than 5% passing no. 200 sieve; and not meeting both criteria for GW
GM	More than 12% passing no. 200 sieve; Atterberg's limits plot below A line (Fig. 7.3) or plasticity index less than 4
GC	More than 12% passing no. 200 sieve; Atterberg's limits plot above A line (Fig. 7.3); and plasticity index greater than 7
GC-GM	More than 12% passing no. 200 sieve; Atterberg's limits fall in hatched area marked CL-ML in Fig. 7.3
GW-GM	Percent passing no. 200 sieve is 5 to 12; and meets the criteria for GW and GM
GW-GC	Percent passing no. 200 sieve is 5 to 12; and meets the criteria for GW and GC
GP-GM	Percent passing no. 200 sieve is 5 to 12; and meets the criteria for GP and GM
GP-GC	Percent passing no. 200 sieve is 5 to 12; and meets the criteria for GP and GC

TABLE 7.3   Unified Classification System—Group Symbols for Sandy Soil
(Reproduced with permission from Ref. 6.)

Group symbol	Criteria
SW	Less than 5% passing no. 200 sieve; $C_u = D_{60}/D_{10}$ greater than or equal to 4; and $C_c = \dfrac{(D_{30})^2}{D_{10} \times D_{60}}$ between 1 and 3
SP	Less than 5% passing no. 200 sieve; and not meeting both criteria for SW
SM	More than 12% passing no. 200 sieve; Atterberg's limits plot below A line (Fig. 7.3) or plasticity index less than 4
SC	More than 12% passing no. 200 sieve; Atterberg's limits plot above A line (Fig. 7.3); and plasticity index greater than 7
SC-SM	More than 12% passing no. 200 sieve; Atterberg's limits fall in hatched area marked CL-ML in Fig. 7.3
SW-SM	Percent passing no. 200 sieve is 5 to 12; and meets the criteria for SW and SM
SW-SC	Percent passing no. 200 sieve is 5 to 12; and meets the criteria for SW and SC
SP-SM	Percent passing no. 200 sieve is 5 to 12; and meets the criteria for SP and SM
SP-SC	Percent passing no. 200 sieve is 5 to 12; and meets the criteria for SP and SC

The classification of fine-grained soils strongly emphasizes plastic properties of the soils which are embodied in the suffix of the group symbol; namely, the letter L denotes low value of liquid limit (below 50 percent), while the letter H is used to describe the high value of liquid limit (above 50 percent).

7.1.4 Dry unit weight (density) and compaction

The density of a soil is one of its most important engineering properties, affecting the behavior of a soil-steel bridge. To enhance its strength and stiffness, the backfill is densified during construction. The factors which influence the density of soil are outlined briefly following, together with the methods for measuring it.

The *density of soil* is usually defined in engineering practice as weight per unit volume (or unit weight). It is different from the more scientific definition of density, which is mass per unit volume. Consistent with current engineering practice, unit weight is also referred to in this book in either grams per cubic centimeter or pounds per cubic foot.

The weight of the dry, solid particles in a unit volume of soil is called the *dry unit weight,* and the weight of the dry solids and the moisture contained in the voids is referred to as the *moist unit weight* or *bulk unit weight. Moisture content* (or *water content*) is defined as the ratio of the weight of water W_w to the weight of solids (soil grains) W_s given in a volume of soil; this ratio in percent is given by:

$$w = \frac{W_w}{W_s} \times 100 \qquad (7.3)$$

The degree of saturation S is formulated as the ratio of the volume of water V_w to the volume of voids V_v, i.e.,

$$S = \frac{V_w}{V_v} \qquad (7.4)$$

For a soil with a given state of compaction, the dry unit weight remains constant; however, the wet unit weight can vary between the limits of dry and saturated unit weights.

The soil unit weight depends mainly upon: (1) the volume of voids relative to the volume of the solid particles; (2) the specific gravity of the solids; and (3) the moisture content. The purpose of compaction is to reduce the volume of voids in the soil, and thus to increase its unit weight. As discussed in several places in the book, highly compacted soil is particularly desirable in the engineered backfill of a soil-steel bridge.

The relative amount of solid matter in a soil may be expressed by means of either the *void ratio* or the *porosity*. The former is defined as the ratio of the volume of voids to volume of solids, and is denoted by e, so that

$$e = \frac{V_e}{V_s} \qquad (7.5)$$

where V_e = the volume of voids in a unit volume of soil and V_s = the volume of solids, also in the unit volume of soil.

The porosity, denoted as a percentage, n, is given by

$$n = \frac{V_e}{V} \times 100 \qquad (7.6)$$

where in a volume V of soil, V_e is the volume of voids. It is noted that both the void ratio and porosity correspond to the total amount of void space, without regard to the amount of moisture contained in the pores.

Denoting the volumes of air voids and moisture as V_a and V_m, respectively, we have

$$V = V_s + V_e \qquad (7.7)$$

and

$$V = V_s + V_a + V_m \qquad (7.8)$$

The percentage of air voids, denoted as n_a, is given by

$$n_a = \frac{V_a}{V} \times 100 \qquad (7.9)$$

Similarly, the percentage of solids denoted as n_s, is given by

$$n_s = \frac{V_s}{V} \times 100 \qquad (7.10)$$

The void ratio e, and porosity n can also be expressed by

$$e = \frac{n}{100 - n} \qquad (7.11)$$

$$n = \frac{100e}{1 + e} \qquad (7.12)$$

It is obvious that

$$n_s + n = 100\% \qquad (7.13)$$

$$n_s + n_a + n_m = 100\% \qquad (7.14)$$

where $n_m = \dfrac{V_m}{V} \times 100$.

Clearly, the higher the void ratio, the smaller the unit weight. Both the unit weight and voids of soil are influenced by the physical state and arrangement of the soil particles. With the help of the example of a jar full of marbles of the same size, shown in Fig. 7.4, it can be appreciated that no compactive effort can densify appreciably a sample if it is composed of particles of the same size. The soil sample can be compacted effectively only if it contains particles of different sizes, with smaller-size particles filling the voids between larger-size particles. A soil is called *well graded* if it has such a distribution of particles of all sizes that the compacted density can approach the maximum possible. The grading curve of such an ideally well-graded soil is parabolic and is referred to, appropriately, as the *ideal grading curve.*

7.1.5 Relative density

One of the measures of the degree of soil compaction of dry soil is the relative density, D_r, given by

$$D_r = \frac{e_{max} - e}{e_{max} - e_{min}} \times 100 \tag{7.15}$$

where e is the void ratio and e_{max} and e_{min} are its maximum and minimum values, respectively, with the former corresponding to the loosest state of the soil and the latter to its densest state.

The value of e_{max} is determined by drying and pulverizing the soil to same-size particles, pouring it slowly and with minimum disturbance through a funnel into a container of known volume, and then weighing it. In this process, the soil must be dry to prevent "bulking." The value of e_{min} is determined after compacting the soil by a drop hammer, tamper, or vibrator.

Employing the definition of any unit weight, the relationship for relative density of dry soils can be expressed as follows:

$$D_r = \left(\frac{\gamma - \gamma_{(min)}}{\gamma_{(max)} - \gamma_{(min)}} \right) \frac{\gamma_{(max)}}{\gamma} \times 100 \tag{7.16}$$

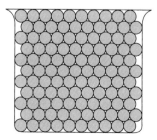

Figure 7.4 A jar full of marbles of the same size.

where $\gamma_{(min)}$ is the dry unit weight of the soil in the loosest state (corresponding to $e_{(max)}$; $\gamma_{(max)}$ is the dry unit weight of the soil in the densest state (corresponding to e_{min}); and γ is the dry unit weight of the soil in situ (corresponding to e).

The qualitative characteristics of soils described by means of relative densities are presented in Table 7.4. The determination of the minimum and maximum dry weight employed in the definition of relative density can be obtained on the basis of ASTM Test Designation D-2049.

Thus, to determine the minimum dry unit weight of the granular soil, the mold of a volume 2830 cm^3 (0.1 ft^3) is used. The granular soil is poured into the mold by means of a funnel equipped with spout of 12.7-mm (0.5 in) diameter in such way that the average height of fall of the granular material is maintained constant and is equal to 25.4 mm (1 in).

In this way γ_{min} can be expressed as

$$\gamma_{min} = \frac{W_s}{V_m} \tag{7.17}$$

where W_s is the weight of granular soil filling the standard mold, and V_m is the volume of the mold.

The next parameter of dry granular soil appearing in the definition of relative density (i.e., γ_{max}) can be determined according to ASTM Test Designation D-2049 by means of vibration. In order to obtain γ_{max}, the additional surcharge of 13.8 kN/m^2 (2 psi) is imposed on the top of granular soil which is placed in the standard mold with volume of 2830 cm^3 (0.1 ft^3). Then the mold with granular soil is subjected to vibration at a frequency of 3600 cycles/min, with amplitude of vibration equal to 0.635 mm (0.025 in) for a total of 8 minutes.

The determination of γ_{max} is obtained after vibration from the known weight and volume of the analyzed granular soil.

Values of D_r range from 0 to 100, with the lowest value indicating the soil in its loosest state, and the highest, its densest state.

TABLE 7.4   Qualitative Description of Granular Soil

Relative density (%)	Description of soil
0–15	Very loose
15–50	Loose
50–70	Medium
70–85	Dense
85–100	Very dense

7.1.6   Proctor dry unit weight

Soil compaction was mainly a trial and error process until the early 1930s, when R. R. Proctor discovered and described the relationship between the soil dry unit weight, moisture content, and compactive effort. Through laboratory testing, he found that with the same compactive effort the dry unit weight of a soil sample increases initially with the increase of the moisture content until the latter reaches a certain value, denoted as the *optimum moisture content*. Any further increase in the moisture content beyond its optimum level causes a reduction in the dry density. This phenomenon is illustrated in Fig. 7.5 in the curve labeled "standard compaction."

In the standard Proctor test (Ref. 5), the soil is compacted in a cylindrical mold 102 mm (4 in) in diameter and about 114 mm (4.5 in) high; it is placed in the mold in three layers, with each layer compacted by 25 blows of a metal tamper weighing 24.5 N (5.5 lb) and having a circular striking face of 51-mm (2 in) diameter. The hammer is allowed to fall freely through a height of 0.305 m (12 in) for each blow. The amount of energy thus applied was established by Proctor as the energy which would give a maximum density in the laboratory test, and which is compatible with the energy that can be imparted realistically in the field.

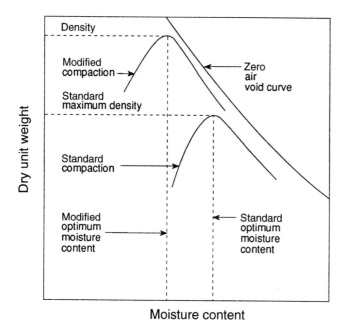

Figure 7.5 Typical Proctor and modified Proctor unit weight curves.

During a standard Proctor test, the energy CE delivered to the soil is equal to:

$$CE = \frac{(3)(25)(24.5)(0.305)}{9.44 \times 10^{-4}(1000)} = 593.7 \text{ kJ/m}^3 \ (12,400 \text{ ft} \cdot \text{lb/ft}^3)$$

The energy imparted to the soil results in:

1. an increase in unit weight
2. an increase in shear strength
3. a decrease in permeability
4. a decrease in shrinkage
5. a decrease in compressibility

The maximum dry unit weight obtained by the standard Proctor test is called the *standard Proctor dry unit weight*, and the moisture content corresponding to this maximum dry unit weight is the *standard optimum moisture content.*

Since it is not always necessary to have field dry unit weight similar to the standard Proctor dry unit weight, specifications for compaction often require that the dry unit weight of the compacted soil achieved in the field should be a certain percentage of the standard Proctor dry unit weight. A specification for relative compaction, R, can be obtained by means of the following relationship:

$$R(\%) = \frac{\gamma d(\text{field})}{\gamma d(\text{max} - \text{lab})} * 100\% \tag{7.18}$$

where $\gamma_{d(\text{field})}$ is the dry unit weight of the granular soil in the field and $\gamma_{d(\text{max} - \text{lab})}$ is the maximum dry unit weight corresponding to optimum moisture content (OMC) obtained in laboratory.

Sometimes, the specifications of the compacted granular soils involve the relative density D_r. In such situations, the following relationship is used for relative compaction:

$$R = \frac{R_0}{1 - D_r(1 - R_0)} \tag{7.19}$$

and

$$R_0 = \frac{\gamma d(\text{min})}{\gamma d(\text{max})} \tag{7.20}$$

Lee and Singh (Ref. 13), based on experimental research, proposed the following relationship for relative compaction R, which also involves relative density D_r,

$$R = 80 + 0.2D_r \tag{7.21}$$

For the engineered backfill of a soil-steel bridge, it is usual to specify that the soil should be compacted to between 85 and 95 percent standard Proctor dry unit weight.

7.1.7 Modified Proctor dry unit weight

The U.S. Army Corps of Engineers has introduced a modification to the standard Proctor test. In this modified test the compactive energy is greatly increased, causing the compacted unit weight to be higher than that achieved in the standard test. The same mold as that used in the standard test is used in the modified test; however, the soil is placed in five layers, the weight of the metal tamper is increased to 44.5 N (10 lb), and the height from which it is dropped is increased to 0.459 m (18 in). The maximum unit weight obtained by the modified method is referred to as the *modified Proctor dry unit weight,* and also as the *modified AASHTO dry unit weight;* the corresponding moisture content is known as the *modified optimum moisture content.*

The energy CE delivered to the soil during the modified Proctor test is equal to:

$$CE = \frac{(5)(25)(44.5)(0.46)}{(9.44)(10^{-4})(1000)} = 2710.5 \text{ kJ/m}^3 \ (56{,}300 \text{ ft} \cdot \text{lb/ft}^3)$$

This compaction energy is about five times larger than that of the standard Proctor test.

For this extra energy, an increase in dry unit weight is around 5 to 10 percent. The compaction curve obtained from the modified Proctor test is always translated to the left and raised up in comparison to the compaction curve obtained from the standard Proctor test (Fig. 7.5). This means that the modified Proctor leads to an increase in the dry unit weight and a decrease in moisture content.

7.1.8 Zero-air-void curve

During the compaction process, the soil starts as a three-phase system which is composed of soil, water, and air. During the initial stage of compaction, which is represented by the dry side of optimum moisture content (OMC), considerable air is present within the soil. This stage of the process results in a change of state with more soil and water being present in the compacted soil. At optimum moisture content there is still a considerable amount of air present in the soil. On the wet side of optimum moisture content (OMC) which is located on the right-hand side of OMC, the main effort of compaction is to replace more and more air with water. If the compaction process were completely efficient, it might be possible to fill all voids with water and obtain a two-phase system. However, it is never possible to obtain such a two-phase system (with degree of saturation $S = 100\%$) as the result of compaction; thus, any curve will always fall below the zero-air-void curve.

Figure 7.5 shows a curve labeled the "zero-air-void curve." This curve represents the dry unit weight of the soil when it is composed only of solid particles and water. Since compaction is the process of eliminating air voids from the soil, this curve represents the upper bound of all the moisture-content–dry-unit-weight curves. As can be seen in Fig. 7.5, an increase in the compactive effort brings the actual curves closer to the zero-air-void curve. Since this latter curve represents a theoretical upper bound of the dry unit weight at a given level of moisture content, it is often shown on moisture versus dry unit weight plots. The zero-air-void unit weight, γ_{dt}, corresponding to any moisture content, is given by

$$\gamma_{dt} = \frac{G\,\gamma_w}{1 + G(w/100)} \tag{7.22}$$

where G is the specific gravity of soil particles, γ_w is the unit weight of water and w is the moisture content as a percentage.

7.1.9 Angle of internal friction

Compacting granular soils to a high unit weight results in an increase of the shear strength, which is connected with the angle of internal friction ϕ. As was mentioned previously, the relative compaction R is related to relative density D_r of granular soils. Consequently, the relative density D_r can be considered as a parameter of other quantities affected by the compaction process. An approximate relationship between relative density and the angle of internal friction is presented in Fig. 7.6. The increase of shear strength obtained from the compaction process of granular soils can be shown by observing the lower bound of the band. The range of variability of ϕ defined by the lower band is contained between 25 and 41°.

The former value of ϕ is associated with the loosest state of dry soil, while the latter is associated with the densest.

7.1.10 Requirements for backfill selection

It cannot be overemphasized that the soil in the engineered backfill around the conduit of a soil-steel bridge must consist of clean, granular, and free-draining material. Such soils possess the desirable elastic and time-independent properties; their shear strength is not affected adversely by the presence of moisture.

In North America, three major documents contain specifications for the design and construction of soil-steel bridges; these documents are:

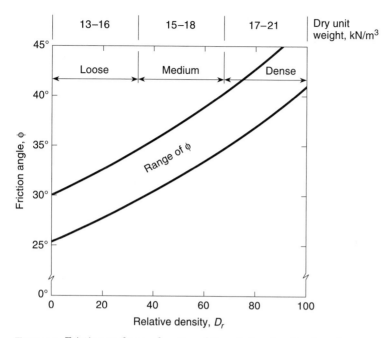

Figure 7.6 Friction angle as a function of the relative density of sand.

(1) the American Iron and Steel Institute (AISI) handbook (Ref. 2); (2) the American Association of State Highway and Transportation Officials (AASHTO) specifications for highway bridges (Ref. 1); and (3) the *Ontario Highway Bridge Design* (OHBD) *Code* (Ref. 17). The requirements for backfill selection given in these documents are given here, mainly to emphasize the importance of backfill in the structural integrity of the bridge.

The AISI handbook requires that the same quality of material be used in the engineered backfill envelope as that customarily used for roadway embankments. In addition it recommends that the material be granular. Cohesive soils are permitted by the AISI for small conduits, provided that the backfill is compacted carefully at the optimum moisture-content level. Very fine granular material is recommended to be avoided when the water table is expected to be high, since in this condition the fine particles can infiltrate into the pipe. Bank-run gravel or similar material compacted to 90 to 95 percent standard density is considered to be ideal by the AISI handbook. Compaction of all soils to at least of 85 percent, standard AASHTO density is considered to be a minimum for all situations.

The classification for granular soils, adopted by the AASHTO specifications, is reproduced in Table 7.5. For all "long-span" structures, which are defined in Chap. 5, Sec. 5.2, the backfill must consist of one of the soils given in Table 7.5 other than soils A-2-6 and A-2-7, which are not considered suitable for soil-steel bridges. When the depth of soil cover is more than 3.66 m (12 ft), only soils A-1 and A-3 are permitted to be used for the backfill. For all structures, the structural backfill is required to be compacted to a minimum of 90 percent, standard AASHTO density.

The 1983 edition of the OHBD code allowed the use of three types of soil, namely, groups I, II, and III; however, it encouraged the use of groups I and II soils (defined in Chap. 5, Table 5.2) in the engineered backfill, and discouraged the use of group III soils, which are fine-grained and with low plasticity. While that code did not explicitly forbid the use of a particular soil in the engineered backfill, it specified a very low value of the modulus of soil stiffness, E', for unsuitable soils. The low values of E' in turn give a small value of buckling strength of the conduit wall, thereby compelling the designer to select either an unusually thick plate or a better-quality soil. This approach has been modified with the 1991 issue of the OHBD code, in which soil group III is eliminated for use as backfill.

7.2   Soil as an Elastic Medium

Despite their nonhomogeneous nature at the microlevel, soils are usually analyzed as elastic media at the macrolevel. The fundamentals of the theory of elasticity applied to soils are presented in this section.

An elastic material is one which regains its original shape completely upon the removal of applied loading. As shown in Fig. 7.7a, the stress-strain (σ–ε) relationship may be linear, in which case the mate-

TABLE 7.5   Classification of Soils According to AASHTO Specifications

General classification	Granular materials (35% or less passing no. 200)						
Group classification	A-1		A-3	A-2			
	A-1a	A-1b		A-2-4	A-2-5	A-2-6	A-2-7
Sieve analysis: Percent passing: No. 10 No. 40 No. 200	50 max 30 max 15 max	— 50 max 25 max	— 51 min 10 max	— — 35 max	— — 35 max	— — 35 max	— — 35 max
Characteristics of fraction passing no. 40: Liquid limit Plasticity index	— 6 max		— NP	40 max 10 max	41 min 10 max	40 max 11 min	41 min 11 min
Usual types of significant constituent materials	Stone fragments, gravel and sand		Fine sand	Silty or clayey gravel and sand			
General rating as backfill material	Excellent to good						

rial is called linear-elastic. When the σ–ε relationship is nonlinear as shown in Fig. 7.7*b*, the material is referred to as nonlinear elastic. It can be seen in Fig. 7.7 that in both the linear and nonlinear cases, the σ–ε curves follow the same path in both the loading and unloading sequences.

One of the simplest methods of analysis of nonlinear elastic material is the simulation of the given stress-strain curve by means of curve-fitting. Suitably, the given stress-strain curve is modeled by means of two straight lines (bilinear material), which can approximately and realistically describe the analyzed material. The behavior of the bilinear material presented in the σ,ε coordinate system is shown in Fig. 7.8.

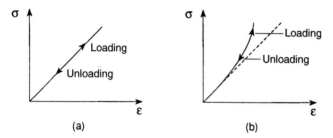

Figure 7.7 Stress-strain relationship of elastic materials: (*a*) linear elastic material; (*b*) nonlinear elastic material.

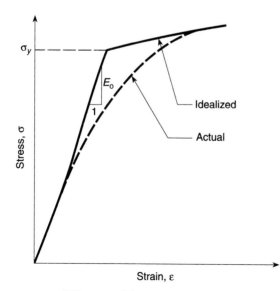

Figure 7.8 Bilinear model for nonlinear material.

Clearly, the initial straight line defines an initial modulus which is valid until the stresses reach a value associated with the yield stress. This point constitutes the starting point of validity of the second straight-line branch, having another constant value of modulus.

7.2.1　Linear-elastic model

The stress-strain relationship of a linear elastic medium is governed by the familiar Hooke's law of elastic deformations (e.g., Ref. 22). In soil mechanics, it is commonly accepted that compressive normal stresses are considered as positive. When the elastic medium is in a state of plane-strain, this relationship can be defined by the following set of three equations:

$$\left.\begin{array}{l} \sigma_x = c_{11}\varepsilon_x + c_{12}\varepsilon_y \\[4pt] \sigma_y = c_{12}\varepsilon_x + c_{22}\sigma_y \\[4pt] \tau_{xy} = c_{33}\gamma_{xy} \end{array}\right\} \tag{7.23}$$

where, as illustrated in Fig. 7.9, σ_x and σ_y are the stresses in the x and y directions, respectively; ε_x and ε_y are their strain counterparts, and τ_{xy} and γ_{xy} are the shear stresses and strains, respectively. If the elastic medium is isotropic, the four constants used in Eqs. (7.23) can each be defined in any one of the three ways, each of which involves only two independent constants. The equations are as follows:

$$\left.\begin{array}{l} c_{11} = c_{22} = \dfrac{E(1-v)}{(1+v)(1-2v)} \\[12pt] \qquad\quad = B + \dfrac{4}{3}G \\[12pt] \qquad\quad = M_s \end{array}\right\} \tag{7.24}$$

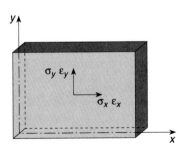

Figure 7.9　Notation for stresses and strains for the state of plane-strain.

$$c_{12} = \frac{E_v}{(1 + v)(1 - 2v)}$$

$$= B - \frac{2}{3}G$$

$$= M_s k_0$$

$$(7.25)$$

$$c_{33} = \frac{E}{2(1 + v)}$$

$$= G$$

$$= \frac{M_s(1 - k_0)}{2}$$

$$(7.26)$$

where E = the modulus of elasticity
v = Poisson's ratio
B = bulk modulus
G = shear modulus
M_s = confined modulus
k_0 = lateral coefficient

It is useful to define the last four constants in terms of the familiar modulus of elasticity and Poisson's ratio.

$$B = \frac{E}{3(1 - 2v)} \qquad (7.27)$$

$$G = \frac{E}{2(1 + v)} \qquad (7.28)$$

$$M_s = \frac{E(1 - v)}{(1 + v)(1 - 2v)} \qquad (7.29)$$

$$k_0 = \frac{v}{1 - v} \qquad (7.30)$$

The confined modulus M_s is also called the *constrained modulus*. It represents the ratio of axial stress to axial strain for confined compression.

Figure 7.10 outlines the geometric representation of the various moduli applied in soil mechanics.

By idealizing the soil as an elastic medium, it is implied that its properties are time-independent.

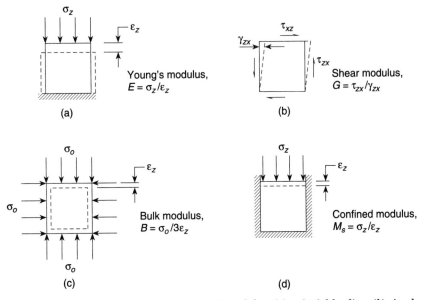

Figure 7.10 Definitions of various types of soil modulus: (a) uniaxial loading; (b) simple shear; (c) isotropic compression; (d) confined compression.

7.2.2 Nonlinear elastic models

Strictly speaking, the stress-strain relationships of even those soils that can be realistically regarded as time-independent are nonlinear. It has been confirmed by many investigators that soils stiffen up with increase in stresses. A typical relationship between stress σ and strain ε is shown in Fig. 7.11a. The upward concavity of the σ–ε curve indicates the stiffening behavior. It is usual to assume that a small increment of σ is related to the corresponding increment in ε in the same way as σ and ε are related for the linear-elastic case. Based on this assumption, the slope of the σ–ε curve at point A, shown in Fig. 7.11a, can be regarded as the elastic modulus of the material corresponding to σ_A which is the stress at point A. This modulus is often referred to as the *tangent modulus*.

It is possible to relate the stress σ_A at point A to the corresponding strain ε_A, by using the slope of the line 0A shown in Fig. 7.11b, where the point 0 corresponds to $\sigma = \varepsilon = 0.0$. The modulus corresponding to 0A is usually referred to as the *secant modulus*.

As shown in Fig. 7.11c, the nonlinear σ–ε curve can be approximated by a number of straight lines, each having its own average value of the tangent modulus. It is usual for the nonlinear analysis of soils to be conducted either by this piecewise linear approach or by the secant modulus approach.

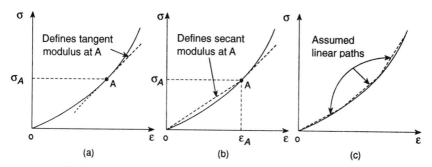

Figure 7.11 Nonlinear stress-strain curves: (*a*) definition of tangent modulus; (*b*) definition of secant modulus; (*c*) approximation by linear-elastic behavior.

7.3 Soil Parameters for Continuum Analysis

Many simplified methods for the force analysis of soil-steel bridges are based upon representing the soil as a linear-elastic medium. Despite its limitations, this approach is considered to be realistic for design purposes. Specific values of the various parameters required for this linear-elastic idealization are discussed in this section.

7.3.1 Poisson's ratio and lateral coefficient

The Poisson's ratio, ν, and the lateral coefficient, k_0, for an elastic medium are interrelated according to Eq. (7.30). It has been shown (e.g., in Ref. 10) that even for nonlinear elastic soils, the values of these constants remain almost unaffected by the level of stress in the environment of confined compression. It is, therefore, reasonable to use single values for these constants for the entire soil medium.

Poisson's ratio ν can be determined from the results of triaxial tests in which volumetric strains of the cylindrical sample are monitored; however, such tests are not routinely performed. For these tests, the ratio of minor and major stresses, σ_3 and σ_1, respectively, can be considered to be approximately equal to the ratio k_0 of the in situ earth pressures, i.e.,

$$k_0 \simeq \frac{\sigma_3}{\sigma 1} \tag{7.31}$$

Several researchers (e.g., Refs. 4 and 8) have suggested the following expression for k_0:

$$k_0 = 1 - \sin \phi \tag{7.32}$$

where ϕ is the angle of internal friction of the soil.

In the absence of test data, ν may be assumed to vary between 0.3 and 0.4 for granular soil. Duncan (Ref. 8), however, recommends a value 0.3 for k_0, which leads to $\nu = 0.43$.

According to Ref. 20, Eq. (7.32) is considered to be correct for loose sand. However, for soil which is subjected to compaction, k_0 is recommended to be evaluated from the following relationship:

$$k_0 = (1 - \sin \phi) + \left[\frac{\gamma d}{\gamma d(\min)} - 1 \right] 5.5 \qquad (7.33)$$

where γd is the dry unit weight of the compacted soil; and $\gamma d(\min)$ is the minimum dry unit weight of the soil in its loosest state.

7.3.2 Secant modulus

The secant modulus E_s of soil depends not only upon the type of soil, but also upon the state of stress.

It is also important to notice that the constrained secant modulus is affected by the cyclic type of loading. In particular, it increases with increasing the number of cycles of loading (Ref. 12).

Some researchers (e.g., Refs. 16 and 21) have suggested that E_s for compacted granular fill can be estimated from

$$E_s = k_s h \qquad (7.34)$$

where h is the depth of the reference point with respect to the top of the fill, and k_s is the constant of soil reaction whose value depends upon the density, or the degree, of the soil. Recommended values of k_s are given in Table 7.6 for different degrees of compaction of the granular soil.

In the backfill of a soil-steel bridge, the state of stress (and consequently E_s) changes from point to point. For simplified analyses, which use a single value of E_s, it is necessary to select a representative value of E_s so that the simplified analysis results are comparable to those obtained by rigorous analyses incorporating the different values of E_s. It has been proposed in Ref. 8 that E_s at a level midway between the crown and spring line, which is identified in Fig. 7.12, is that representative single value. Clearly, this representative value of E_s depends upon the depth of cover and the type and degree of compaction of the

TABLE 7.6   Values of k_s for Granular Soil

Relative density	Percentage of standard Proctor density	k_s lb/in^3	k_s kN/m^3
Loose	<90	<4.0	<1085
Medium	90–100	4.0–12.0	1085 → 3257
Dense	>100	>12.0	>3257

soil. Ref. 8 provides curves for the representative values of E_s for different soils, plotted against h, which (as shown in Fig. 7.12) is the distance of the level midway between the crown and spring line from the top of the fill. These curves are reproduced in Fig. 7.13. The notation used in this figure for soil classification is the same as that defined in Tables 5.2 (Chap. 5), 7.2, and 7.3.

7.4 Parameters for Soil Resistance

For the analysis of the effect of foundations on the soil, it is usual to idealize the soil as a series of springs, and to represent the properties of these springs by parameters such as:

- modulus of soil stiffness, E'
- coefficient of normal reaction, k_n
- coefficient of shear reaction, k_s

These parameters have been found to be helpful in the analysis of several problems of soil-structure interaction, including the buckling of conduit walls. It should be noted that these parameters are based upon the following simplifying assumptions:

1. The displacement at a given point on the soil-structure interface is assumed to be directly proportional to the pressure developed at that point; this assumption is not strictly valid since the stiffness of a soil, as discussed in Sec. 7.2, depends upon the level of stress.

2. E', k_n, and k_s are all often assumed to be unique properties of the soil medium, whereas these parameters (as shown, for example, in Refs. 18 and 19), also depend upon the type of structure, the loading configuration, the relative stiffness of the structure, and the response under investigation.

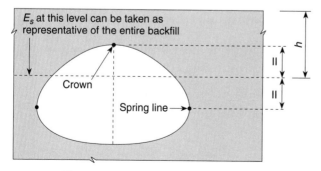

Figure 7.12 Identification of reference level for representative E_s.

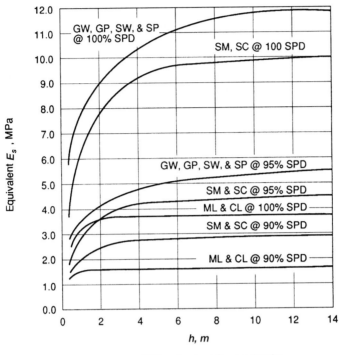

NOTE: SPD = standard Proctor density.
(For definition of *h* see Fig. 7.12.)

Figure 7.13   Equivalent values of E_s.

Within these limitations, the values of the parameters under consideration can be determined analytically or experimentally through a reverse process. In this process, a solution, or formula, is first developed based on one of these parameters, and the results of the problem are obtained either analytically or experimentally. The value of the parameter is then determined so that, when substituted in the already developed solution or formula, it would yield the same result as that obtained analytically or experimentally. This process will become clearer when discussed subsequently in the context of determining the modulus of soil stiffness E'.

7.4.1   Modulus of soil stiffness

The modulus of soil stiffness, E', represents the stiffness of the soil at the soil-pipe interface and is defined as interface pressure divided by soil strain.

$$E' = \frac{p}{\varepsilon} \qquad (7.35)$$

where p is the radial pressure at the interface and ε is the radial strain, also at the interface. E' has the units of stress.

As discussed in Chap. 4, Sec. 4.5, the crown deflection of a pipe embedded in soil is related to the flexural rigidity of the pipe and E', as shown in Eq. (4.35), which is also known as the Iowa formula. Using this equation as the basis, the values of E' were back-calculated for different soils by Howard (Ref. 9), using experimental values of the crown deflection Δ_x. The values of E' proposed in this reference are reproduced in Table 7.7; the notation used in this table for defining the various type of soils is defined in Chap. 5, Table 5.2.

It is noted that E' is sometimes referred to as the *modulus of soil reaction*. As mentioned in Ref. 9, E' is not simply a property of the soil, although it does depend predominantly on the type of soil and the degree of its compaction.

The relationship between E' and a constant secant modulus E_s of the soil can be established by considering a plate of infinite extent with a circular hole and in a state of plane-strain (Refs. 11, 22). The plate is assumed to be of a linear-elastic material, with modulus of elasticity E_s and Poisson's ratio v, and is subjected to a radial pressure p acting outwardly in the hole, as shown in Fig. 7.14. It can be shown that the radial strain ε_r in the plate at a distance R_i from the center of the hole is given by

$$\varepsilon_r = \frac{p\,R^2}{E_s R_i^2}(1 + v) \qquad (7.36)$$

where R is the radius of the hole. By putting $R_i = R$ in Eq. (7.36) and denoting ε_r as ε, Eq. (7.34) gives

$$E' = \frac{E_s}{1 + v} \qquad (7.37)$$

TABLE 7.7   Empirical Values of E' Proposed in Ref. 9

Soil type	E' in MPa (psi) for Proctor density		
	<85%	85–95%	>95%
Fine-grained soils, CL, ML, ML-CL with <25% coarse-grained particles	1.4 (200)	2.8 (400)	6.9 (1000)
Fine-grained soils, CL, ML, ML-CL with <25% coarse-grained particles	2.8 (400)	6.9 (1000)	13.8 (2000)
Coarse-grained soils with fixes, GM, GC, SW, SC	2.8 (400)	6.9 (1000)	13.8 (2000)
Coarse-grained soils with few or no fixes, GW, GP, SW, SP	6.9 (1000)	13.8 (2000)	20.7 (3000)

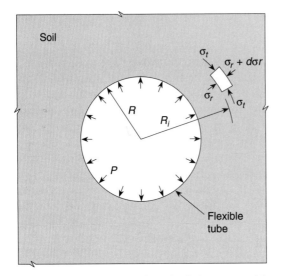

Figure 7.14 A plane-strain slice of infinite extent with a circular hole.

Strictly speaking, Eq. (7.37) is valid only when the plate is of infinite extent. Reference 14 contains details of a study in which the expression for E' was calculated for the same case as that shown in Fig. 7.14, but in which the plate of infinite extent was replaced by a soil ring of finite thickness, as shown in Fig. 7.15. As shown in this figure, the outer boundary of the soil ring is assumed to be contained by a rubber membrane. Reference 14 provides the following expression for E' for this case:

$$E' = \frac{[1 - (R/R_1)^2]\, E_s}{(1 + \nu)\{1 + (R/R_1)^2\,(1 - 2\nu)\}} \tag{7.38}$$

where R_1 is the outer radius of the soil ring and R, the inner radius. It is interesting to note that as R_1 tends to infinity (i.e., when the soil tends to infinite extent), Eq. (7.38) becomes the same as Eq. (7.37).

Some researchers (e.g., Refs. 15 and 16) have proposed the following expression for the approximate value of E':

$$E' = \frac{E_s}{2(1 - \nu^2)} \tag{7.39}$$

It has been suggested in Refs. 15 and 16 that, for circular tubes having shallow depths of cover, the effective modulus of soil reaction, denoted as E''', can be obtained from

$$E'''_m = E'\left[1 - \left\{\frac{R}{R+h}\right\}^2\right] \tag{7.40}$$

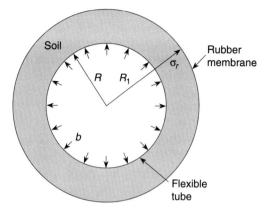

Figure 7.15 A ring of soil in the state of plane-strain.

where h is the average depth of cover and E' is the modulus corresponding to a deeply buried conduit.

7.4.2 Coefficient of normal reaction

The coefficient of normal reaction, k_n, is defined as the normal pressure divided by the displacement at the soil-pipe interface; it has the units of force/length^3, e.g., kN/m^3 or lb/in^3.

$$k_n = \frac{p}{\Delta_n} \tag{7.41}$$

where p is the radial pressure at the soil-structure interface and Δ_n is the radial displacement also at the interface.

An expression for k_n for a circular pipe buried in an elastic medium of infinite extent can be developed in the same way that Eq. (7.37) was developed for E' earlier, with reference to Fig. 7.14. The radial displacement of the soil-pipe interface of the case shown in this figure can be obtained by integrating the radial strains from $R_i = R$ to ∞. Using Eq. (7.36), this integration gives

$$\Delta_n = \frac{Rp}{E_s(1 + v)} \tag{7.42}$$

Substituting this expression for Δ_n in Eq. (7.41) gives

$$k_n = \frac{E_s}{R(1 + v)} \tag{7.43}$$

or, using Eq. (7.37),

$$k_n = \frac{E'}{R} \tag{7.44}$$

Strictly speaking, Eqs. (7.43) and (7.44) are valid for only circular conduits. However, since similar solutions are not available for noncircular conduits, it is suggested in Ref. 11 that these expressions can also be used for noncircular conduits by regarding R as half the conduit span. However, since Δ_n is calculated as the integral of strains starting from the face of the conduit, the authors of this reference recommend that for noncircular conduits, the local radii should be used when calculating k_n from Eq. (7.44).

Through a study of the problem illustrated in Fig. 7.15, Ref. 14 suggests the following expression for k_n

$$k_n = \frac{E_s\{1 - (R/R_1)^2\}}{R(1 + v)\{1 + (R/R_1)^2(1 - 2v)\}} \tag{7.45}$$

It is noted that Eq. (7.45) becomes the same as Eq. (7.43) when $R_1 \to \infty$.

Following the approximate Eq. (7.39) proposed in Refs. 15 and 16, the expression for k_n can also be written as

$$k_n = \frac{E_s}{2R(1 - v^2)} \tag{7.46}$$

Recognizing that k_n is not exclusively a soil parameter, an intensive study reported in Ref. 18 was undertaken to develop simple expressions for this parameter, taking account of all the relevant properties of soil-steel bridges. The study was conducted by relating the results of plane-strain, finite element analyses to those obtained by analyzing the structures as plane-frames supported on springs. The latter method of analysis, which explicitly utilizes k_n, is described in Chap. 4, Sec. 4.4. The study under consideration led to the following expression for k_n:

$$k_n = \gamma \beta\, C_d\, C_\theta\, \sqrt{H/D_h} \tag{7.47}$$

where

$$C_d = 4.25 - \frac{0.75D_h}{100} \tag{7.48}$$

$$C_\theta = 0.25\left(1 + \frac{5.4\theta}{\pi}\right) \tag{7.49}$$

$$\left.\begin{array}{ll} \beta = 1.0 & \text{for dense compacted fill} \\[2mm] \beta = 0.45 + \dfrac{D_h}{200}\left\{\dfrac{\theta}{\pi} - 0.5\right\} & \text{for medium-dense compacted fill} \end{array}\right\} \tag{7.50}$$

in which γ = unit weight of soil in lb/in³
 H = depth of fill at the point where k_n is computed, in inches
 D_h = span of the conduit in inches
 θ = the angular coordinate in radians with the coordinate system illustrated in Fig. 7.16.

It is emphasized that the preceding expressions are not unit-independent and that they use quantities in U.S. customary system of units. The following conversion factors may be found useful if the basic data is in metric units:

$$1.0 \text{ in} = 0.0254 \text{ m}$$

$$1.0 \text{ lb/in}^3 = 0.2714 \text{ MN/m}^3$$

The influence of the relative position of the point of investigation on the value of k_n can be seen in Fig. 7.17. This figure shows the values of k_n plotted around the conduit for one particular case in which the conduit is round and the depth of cover rather shallow. It can be seen in this figure that the value of k_n at the crown is about one-tenth that at the invert.

7.4.3 Coefficient of tangential reaction

The coefficient of tangential reaction, k_s, is defined as the shear force divided by the tangential displacement at the soil-pipe interface, these being denoted by v and Δ_s, respectively.

$$k_s = \frac{v}{\Delta_s} \tag{7.51}$$

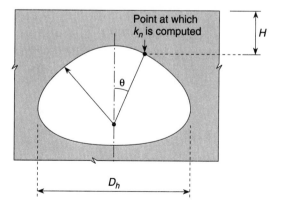

Figure 7.16 Illustration of notation for calculating k_n.

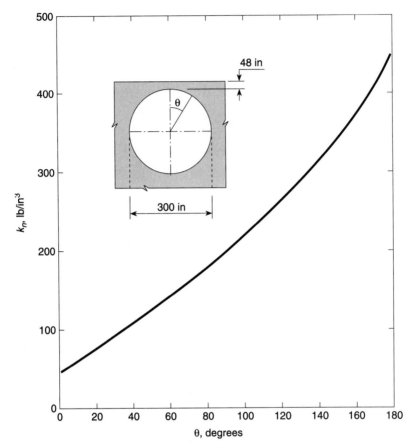

Figure 7.17 Variation of k_n around the conduit for one particular case.

This coefficient has not been dealt with extensively in recent technical literature. However, Ref. 19 shows that for all practical purposes, it is independent of the location of the point of investigation and its value is about 20 percent of the value of k_n at the invert.

References

1. American Association of State Highway and Transportation Officials (AASHTO), *Standard Specification for Highway Materials,* 10th ed., part I, Washington, D.C. 1970.
2. American Iron and Steel Institute, *Handbook of Steel Drainage and Highway Construction Products,* New York, N.Y., 1971.
3. ASTM Committee D-18, *Procedure for Testing Soils,* 4th ed., Amer. Soc. for Testing and Materials, Philadelphia, Pennsylvania, 1964.

4. Bishop, A. W., *The Strength of Soils as Engineering Materials,* Geotechnique, June, 1966, pp. 89–130.
5. Bowles, J. E., *Engineering Properties of Soils and Their Measurements,* 3d ed., McGraw-Hill Book Company, 1986.
6. Das, B. M., *Principals of Geotechnical Engineering,* 2d edition, PWS-KENT Publishing Company, Boston, 1990.
7. Desai, C. S., and Christian, J. T., *Numerical Methods in Geotechnical Engineering,* McGraw-Hill Book Company, 1977.
8. Duncan, J. M., "Behaviour and Design of Long Span Metal Culverts," *Journal of the Geotechnical Engineering Division,* ASCE, 105 (GT3) March, 1979.
9. Howard, A. K., "Modulus of Soil Reaction (*E'*) Values for Buried Flexible Pipe," U.S. Department of Interior, Bureau of Reclamation, Engineering Research Center, Denver, Colo., 1976.
10. Katona, M. G., and Smith, J. M., CANDE, *Users Manual,* Civil Engineering Laboratory, Naval Cons. Battalion Center, Port Hueneme, Calif., 1976.
11. Kloeppel, K., and Glock, D., "Theoretische und experimentelle Untersuchungen zu den Traglastproblemen beigeweicher, in die Erde eingebetter Rohre," no. 10 publication, Institute Für Statik und Stahlbau, T. H. Darmstadt, Germany, 1970.
12. Lambe, T. W., and Whitman, R. V., *Soil Mechanics,* John Wiley & Sons, 1969.
13. Lee, K. W., and Singh, A., "Relative Density and Relative Compaction," *Journal of the Soil Mechanics and Foundations Division,* ASCE, vol. 97, no. SM7, pp. 1049–1052.
14. Luscher, U., "Buckling of Soil-surrounded Tubes," *ASCE Journal of the Soil Mechanics and Foundation Division,* pp. 211–228, 1966.
15. Meyerhof, G. G., and Baikie, L. D., "Strength of Steel Culvert Sheets Bearing against Compacted Sand Backfill," Highway Research Board, *Proceedings of the Annual Meeting,* 42, 1963.
16. Meyerhof, G. G., "Some Problems in the Design of Shallow-buried Steel Structures," *Proceeding of Canadian Structural Engineering Conference,* Toronto, Ont., 1968.
17. Ministry of Transportation of Ontario, *Ontario Highway Bridge Design Code,* Downsview, Ontario, Canada, 1991.
18. Okeagu, B. N., "Analysis and Stability of Large-Span Flexible Conduits," Ph.D. dissertation, Michigan State Univ., 1982.
19. Okeagu, B. N., and Abdel-Sayed, George, "Coefficient of Soil Reaction for Buried Flexible Conduits," *ASCE J. Geotech. Eng.,* 11(7), July, 1984, pp. 908–922.
20. Sherif, M. A., Fang, Y. S., and Sherif, R. I., "K_A and K_0 Behind Rotating and Non-Yielding Walls," *Journal of Geotechnical Engineering,* ASCE, vol. 110, no. GT7, pp. 41–56.
21. Terzaghi, K., "Evaluation of Coefficient of Subgrade Reaction," *Geotechnique,* 5, 1955, p. 297.
22. Timoshenko, S. P., and Goodier, J. N., *Theory of Elasticity,* 3d ed., McGraw-Hill Book Company, 1970.

Special Features

Baidar Bakht

Ministry of Transportation of Ontario
Downsview, Ontario, Canada

George Abdel-Sayed

University of Windsor
Windsor, Ontario, Canada

In its simplest form, a soil-steel bridge contains no other structural elements than the compacted backfill and the metallic shell made only out of lapped, corrugated steel plates. This simple structure, even with plates of the largest available thickness being 7.0 mm (0.28 in), cannot usually have spans larger than about 9.0 m (30 ft). Larger spans are possible only if, by some means, the load effects in the conduit wall are reduced or its load carrying capacity is increased; these means are referred to as "special features" for soil-steel bridges, and are the subject of this chapter.

Depending upon the dominant manner in which they enhance the load-carrying capacity of a soil-steel bridge, the various special features can be grouped into the following three categories:

1. Features which reduce load effects, in particular thrust, in the conduit wall

2. Features which increase the strength of the conduit wall by reinforcing it

3. Features which increase the strength of the conduit wall by stiffening the soil and thus enhancing the stiffness of the radial support to the conduit wall

Special features grouped in these three categories are discussed under separate headings in the following sections. Rigorous design methods dealing with many of these features are not given in this book because they are not well established and, at present, are not generally available. Approximate techniques for dealing with several of the special features are, however, presented.

8.1 Special Features for Reducing Conduit Wall Thrust

The dead-load thrust in the conduit wall can be reduced by creating conditions which promote positive arching through the kind of behavior illustrated in Chap. 3, Fig. 3.7b. It is recalled that positive arching occurs when the column of soil directly above the conduit deflects downward with respect to the adjacent columns of soil.

Live-load thrust in the conduit can be reduced by stiffening the medium above the conduit so that the concentrated loads applied at the top of the fill disperse to greater areas at the crown level, and thereby induce smaller pressures on the pipe.

It will be appreciated that the special features for reducing the dead-load thrust are useful when the fill above the conduit is relatively deep. On the other hand, special features for reducing the live-load thrust are particularly useful when the depth of soil cover is small, because in this case the live-load thrust constitutes a significant fraction of the total thrust in the conduit wall.

Mechanics of arching.   Before discussing the various special features which induce positive arching, it is appropriate to discuss the mechanics of arching as generated by an induced downward movement of the column of soil directly above the conduit. An arch structure in which the foundations are supported by conceptual springs is selected as a vehicle for this discussion. The cross section of such a completed structure is shown in Fig. 8.1a in both its initial and deflected positions.

Figure 8.1b shows a stage during construction when the fill has been raised above the conduit and a layer of loose backfill has just been added to the layers already compacted. It can be appreciated readily that, while this latest layer of backfill will cause the foundation to settle somewhat, its own load will be transferred to the conduit wall without the benefit of positive arching induced by the foundation settlement. This is because this layer of backfill, lacking compaction, does not have enough stiffness to participate in the arching action. However, the foundation settlements caused by the addition of this fill will help to relieve the conduit wall thrust due to the compacted layers of fill below it.

If the springs have a linear load-deformation, or T-Δ, relationship (as shown in Fig. 8.2a), then clearly their settlement during the placement of earlier lifts of backfill above the conduit adds little to the develop-

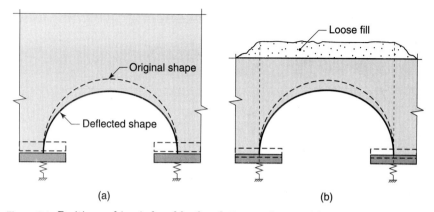

Figure 8.1 Positive arching induced by foundation settlement: (*a*) cross section of completed structure; (*b*) cross section of structure during construction.

ment of the arching action. An ideal spring would be one which is nondeflecting in the initial stages of loading, and which exhibits a softening behavior for higher levels of loading. Such an ideal load-deformation curve is shown in Fig. 8.2*b*, together with a corresponding desirable one which is easier to attain in practice.

8.1.1 Slotted bolt holes

It can be readily appreciated that if the axial stiffness of the conduit wall is very low, the circumference of the pipe will shorten under dead loads, as is illustrated in an exaggerated manner in Fig. 8.3. This reduction in circumference will generate positive arching by causing the column of soil above the conduit to settle down, as is also illustrated in Fig. 8.3.

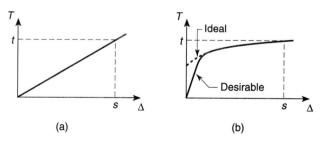

Figure 8.2 Load-displacement characteristics of foundation support: (*a*) linear; (*b*) nonlinear.

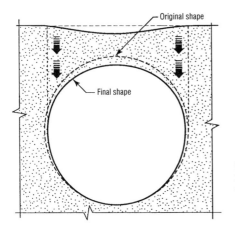

Figure 8.3 Arching induced by reducing the axial stiffness of conduit wall.

The axial stiffness of a typical lap-jointed pipe of a soil-steel bridge is so high in comparison with the relevant stiffness of the surrounding soil that the axial deformations of the pipe are almost negligible. The usual lap joints at longitudinal seams, incorporating circular bolt holes and high-friction bolts, also do not permit enough slippage to induce appreciable arching. A novel method of increasing the axial pipe deformations, and thereby inducing positive arching, is to let the joints slip through bolt holes that are in the shape of keyholes. These holes, which are referred to as "slotted holes," permit slippage of the mating plates that is large enough to generate a substantial degree of positive arching under deep fills. A joint with slotted bolt holes is shown in Fig. 8.4a, and the dimensions of a slotted hole in Fig. 8.4b.

The typical load-slip behavior in compression of a joint with slotted holes is compared in Fig. 8.5 with the load-slip behavior of a joint having circular bolt holes (Ref. 9). The two curves shown in this figure cor-

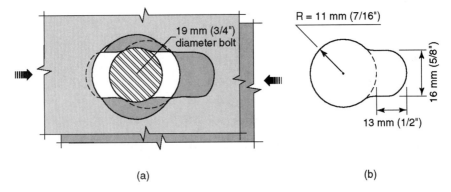

(a)   (b)

Figure 8.4 Slotted joint: (a) jointed plates with section through bolt; (b) dimensions of the keyhole.

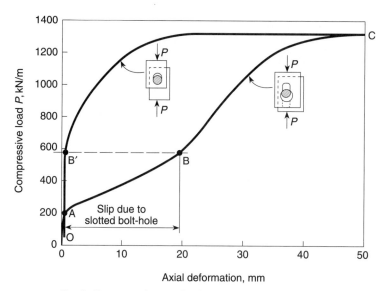

Figure 8.5 Load-slip curves for standard joints and slotted bolt-hole joints for 4.7-mm-thick plates.

respond to plates of nominal thickness of 4.7 mm (0.185 in) and a bolt torque of 270 N·m (200 ft·lb).

It can be seen in Fig. 8.5 that the initial portion of the load-slip curve, identified as OA, is linear and is the same for both the joints; this portion represents the elastic deformation of the material around the bolt holes. When the load is increased beyond A, the joint with circular holes continues to deform in a linear elastic manner until the load reaches point B′, after which level the load-slip behavior becomes markedly nonlinear, leading to the failure of the joint. This last phase of the load-slip behavior of a joint with circular holes is represented by curve B′ C in Figure 8.5.

For the joint with slotted holes, the load beyond point A forces the bolt into the narrow portions of the slotted holes. As expected, the load-slip curve for this joint, represented as AB in Fig. 8.5, follows a much flatter slope than the curve in the earlier phase. The portions of one of the jointed plates that are pushed aside by the bolt traveling from A to B are shown hatched in Fig. 8.6. After the bolt has reached the end of the narrow portion of the slotted hole, the load-slip behavior of the joint becomes almost the same as the load-slip behavior of the joint with circular bolt holes beyond point B′. It can be seen in Fig. 8.5 that the load-slip curve BC for the former joint is very similar to the curve B′ C for the latter. In fact, as observed in Ref. 9, the failure loads for the two joints are practically the same. The implication of this observation

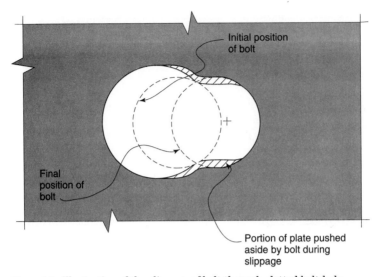

Initial position
of bolt

Final
position of
bolt

Portion of plate pushed
aside by bolt during
slippage

Figure 8.6 Illustration of the slippage of bolt through slotted bolt-hole.

is that, for strength design, the joint with slotted holes can be treated in the same way as the corresponding joint with circular holes.

From the preceding discussion, it follows that the only difference in the load-slip behavior of the joints with slotted and circular holes is the portion represented by curve AB in Fig. 8.5. Given a sufficient number of slotted joints around the circumference of the pipe, the amount of joint slip represented by curve AB in Fig. 8.5 can be significant enough to induce substantial positive arching.

Reference 9 has provided load-deformation characteristics of corrugated plates lap-jointed through slotted bolt holes; these curves, which correspond to plates with 152×51 mm (6×2 in) corrugation profiles and three different thickness, are reproduced in Fig. 8.7. This figure also shows the limits of the deformations, which can be regarded as the end of the slip discussed earlier. As expected, the extent of the total slip does not vary significantly with the thickness of the mating plates. For all practical purposes, the total slip can safely be assumed to be equal to 19 mm (0.75 in). The only portions of the curves of Fig. 8.7 that are likely to be of direct interest to the designers are those that lie between axial deformation of 0 and 19 mm (0.75 in). By using these portions of the curves and interpolating between them visually, a reasonable estimate can be made of the slip that will occur in a joint having two plates each of a given thickness and subjected to a certain axial force.

It can be confirmed from Fig. 5.9 in Chap. 5 that the loads causing failure in conventional joints are considerable higher than the loads which cause the maximum slip in corresponding joints with slotted bolt holes.

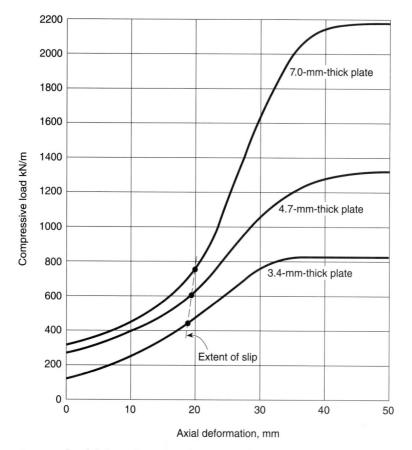

Figure 8.7 Load-deformation curves for joints with slotted bolt-holes.

A pipe with a uniform area of cross section A per unit length is considered. The circumference of the pipe is denoted by C and its shortening due to joint slippage by d_c. The ratio of d_c to C is denoted by the fraction f, so that

$$f = d_c/C \qquad (8.1)$$

Let this pipe be subjected to a uniform thrust T after the slippage of the joints. Using relatively elementary methods, it can be demonstrated that, on the basis of axial deformations, the pipe with joint slip fraction f and cross-sectional area A is equivalent to a pipe with no joint slip and a reduced area of cross section A_r which is given by

$$A_r = A \left\{ \frac{1}{1 + fAE/T} \right\} \qquad (8.2)$$

where E is the modulus of elasticity of the plate steel.

By using A_r instead of A in the calculation of C_s by Eq. (4.23) and then using Eq. (4.22), an estimate can be made of the dead-load thrust in the conduit wall.

It must be emphasized that the interaction between joint slippage and dead-load thrust in the conduit wall is highly nonlinear and complex. This interaction can be accounted for in only an approximate manner by using Eq. (8.2), with the following steps of calculation:

1. From Fig. 8.7, find the value of the thrust that will induce the limiting slip; let this thrust be denoted as T_s.

2. Find the dead-load thrust, T_i, in the conduit wall by ignoring the slippage of joints.

3. If T_i is less than T_s, then use T_i to find slippage per joint, or axial deformation, from Fig. 8.7. When T_i is greater than T_s, then assume a slippage of 19 mm (0.75 in) per joint. Multiply the slippage thus obtained by the number of slotted hole connections around the conduit circumference to obtain d_c, and thence the joint slip fraction f from Eq. (8.1).

4. Use the value of f just calculated and substitute $T = T_i$ to obtain the equivalent area A_r from Eq. (8.2).

5. Use this equivalent area to calculate the joint stiffness parameter C_s from

$$C_s = E_s \frac{D_v}{EA_r} \qquad (8.3)$$

where the notation, other than for A_r, is as defined in Chap. 4 in conjunction with Eq. (4.23).

6. Using the value of C_s just obtained, calculate the thrust reduction, due to joint slippage, δT, from

$$\delta T = 0.1 \times C_s \times (\text{lesser of } T_i \text{ and } T_s) \qquad (8.4)$$

7. Obtain a revised estimate of T_i by subtracting δT from the value obtained in step 2.

Since joint slippage has the effect of reducing the dead-load thrust in the conduit wall, the load factor that is applied to the loads which induce this slippage should have the minimum, rather than the maximum, value. It is recalled, for example, that the *Ontario Highway Bridge Design Code* (Ref. 14) specifies the maximum and minimum load factors for the weight of soil to be 1.25 and 0.80, respectively. The minimum load factor should be used in calculating T_i in step 2, and consequently f in step 3. Of course, the maximum load factor should be used in obtaining the dead-load thrust from T_i by step 7.

The method presented here should be regarded as nothing more than an approximation. Reference 9 should be consulted for further information on the concept of slotted bolt holes.

8.1.2 Squeeze blocks

In the case of arch structures, the metallic shell, and hence the column of soil directly above it, can be made to undergo extra vertical deflection by inserting blocks of a compressible material in the footings of the conduit wall. This concept has been incorporated in a deeply buried arch structure described in Ref. 11, the cross section of which is shown in Fig. 8.8. As shown in this figure, the span and rise of the conduit are 15.5 m (51 ft) and 7.9 m (26 ft), respectively, and the depth of cover is 13.4 m (44 ft). As is also shown in this figure, the structure has two additional special features, namely, transverse stiffeners and longitudinal thrust beams, which are discussed in sections 8.2.1 and 8.2.4, respectively.

The structure under consideration has footings of two different sizes, the smaller footings being located near the ends of the pipe where the depth of soil cover is much smaller than in the middle regions. Details of the larger footing are given in Fig. 8.9, from which it can be seen that the conduit wall is connected to a small concrete footing which transmits its thrust to the larger footing through wooden blocks. The larger footing is of reinforced concrete and the wooden blocks are 0.61 m (2 ft) long in the direction of the conduit span and have a cross section of 305 × 200 mm (12 × 8 in). The wooden blocks have a clear spacing which varies between 100 mm (4 in) at the middle of the pipe and 410 mm (16 in) near the ends.

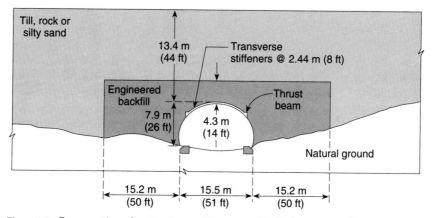

Figure 8.8 Cross section of a structure with squeeze blocks incorporated in the foundations.

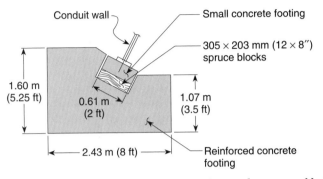

Figure 8.9 Details of the footing incorporating wooden squeeze blocks.

Reference 11, which describes the aforementioned structure and presents extensive test data, claims that the provision of squeeze blocks reduced the dead-load thrust in the conduit wall substantially. However, a quantitative measure of the thrust reduction is not provided in this reference. Notwithstanding the lack of this information, it can be readily appreciated that the use of wood blocks, loaded perpendicular to their grains, as squeeze blocks can be quite effective in promoting positive arching.

Standard textbooks on timber engineering (e.g., Refs. 8 and 12) show that the load-deformation characteristic of wood loaded perpendicular to its grains is as shown conceptually in Fig. 8.10a. As can be seen in this figure, the wooden block softens after the stress perpendicular to the grain exceeds a certain limit. This nonlinear softening behavior is more effective than a linear behavior, as discussed earlier. To take advantage of this softening behavior of wood, however, it is important that the stresses induced in the wood blocks be beyond the proportionality range.

An additional and significant advantage of stressing the wooden squeeze blocks to a high level is that under high, sustained perpendicular-to-grain stress, the deformations increase substantially with time as illustrated in Fig. 8.10b by the curve labeled "high stress." This curve shows that the time-dependent deformations of wood falls into three distinct zones: (1) initially, the deformations increase rapidly with time; (2) the deformations then grow very slowly with time; and (3) after a certain time has elapsed, the deformations again increase very rapidly, leading to a complete collapse of the internal structure of wood. The other curve of Fig. 8.10b, labeled "low stress," shows that under a low level of perpendicular-to-grain stress, the time-dependent deformations virtually do not grow at all from their initial value.

Since the degree of positive arching is related directly to the foundation settlement, and hence to the deformation of the squeeze blocks, it is preferable to let these blocks deform as much as possible. It follows, therefore, that it is preferable to let the conduit wall induce a high enough stress in them to precipitate the kind of long-term behavior illustrated in Fig. 8.5*b* by the curve labeled "high stress."

The effect of the vertical settlement of the pipe which is induced by the deformations of the squeeze blocks can be accounted for approximately in the dead-load analysis through the use of an equivalent reduced axial stiffness of the conduit wall, as was discussed earlier in the context of slotted bolt holes.

8.1.3 Relieving slabs

A relieving slab is a horizontal, or nearly horizontal, reinforced concrete slab located in the backfill of a soil-steel bridge above its conduit. Such slabs are located at or below the embankment level. Details of a soil-steel bridge with a relieving slab at the embankment level are shown in Chap. 1, Fig. 1.30.

Relieving slabs are particularly useful in structures with relatively long spans and shallow soil covers above the crown. In such structures, they serve two functions, i.e., that of reducing the load effects in the pipe due to live loads and that of reinforcing the soil cover above the conduit.

The live-load effects in the conduit walls are reduced by the relieving slab because it affects a much greater dispersion of the concentrated loads through the soil below, thereby reducing the resulting radial pressure on the pipe. Field testing of the structure with relieving slab

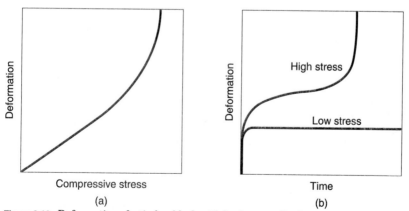

Figure 8.10 Deformation of a timber block with load perpendicular to grain: (*a*) deformations plotted against stress; (*b*) deformations under constant stress plotted against time.

shown in Fig. 1.30 and of a similar structure without the relieving slab, has shown that the presence of the relieving slab can reduce the live-load effects in the conduit wall by up to 50 percent (Ref. 2).

A rigorous method of estimating the reduction of live-load effects in the conduit wall due to the presence of the relieving slab is not available. In the absence of such a method it is recommended that, for the purpose of live-load analysis, the concrete slab be replaced by an equivalent much deeper earth fill which would have the same flexural rigidity as that of the slab. If the thickness of the concrete slab is t_c and the ratio of the moduli of elasticity of concrete and soil is n, then clearly the thickness of the equivalent soil = $n^{1/3}t_c$.

The action of the relieving slab in strengthening the soil cover above the crown is relevant both to the failure of this cover through tension and to a sliding-wedge type of behavior. These two types of soil cover failure are illustrated in Figs. 4.6b and a, respectively. Protection against such failures is provided by code provisions for minimum depth of cover. For example, the *Ontario Highway Bridge Design Code* (Ref. 14) requires that the depth of cover in a soil-steel bridge be not less than $(D_h/6)$ $(D_h/D_v)^{0.5}$, $0.4(D_h/D_v)^2$, or 0.6 m, where D_h and D_v are as defined in Chap. 4, Fig. 4.23. The relieving slab may be found useful in circumstances in which the minimum depth-of-cover requirements of the code cannot be met because of restricted depth of construction. For such applications, a relieving slab can be considered to be equivalent to a much deeper layer of soil having the same resistance to shear as the unreinforced concrete slab.

When the relieving slab is at the embankment level, as in the example in Fig. 1.30, its reinforcement can be designed by carrying out an analysis of a slab on grade. When the slab carries earthfill above it, it may only be necessary to reinforce it with that minimum amount of reinforcement which is required for the control of cracking due to the effects of shrinkage and temperature.

8.2   Special Features Which Reinforce the Conduit Wall

The load-carrying capacity of the metallic shell of soil-steel bridges can be enhanced by attaching appendages to it. Some of the commonly used appendages are discussed in this section.

8.2.1   Transverse stiffeners

The conduit walls of soil-steel bridges having relatively long spans are often stiffened by circumferential stiffeners applied to the top portion of the pipe; these stiffeners are referred to herein as *transverse stiffeners*. The transverse stiffeners may consist of corrugated steel plates of

narrow widths, having the same radius of curvature as the top segments of the pipe, and placed in a ridge-over-ridge fashion, as illustrated in Fig. 8.11a. These stiffeners are spaced at regular intervals, as
can be seen in Fig. 8.12a, which shows the photograph of the metallic
shell of a soil-steel bridge during construction.

As an alternative to corrugated plate stiffeners, there are transverse
stiffeners consisting of curved, rolled components with sections as
shown in Fig. 8.11b, c, d. These latter stiffeners are used with pipes of
very large spans. A pipe with I-section stiffeners can be seen during
construction in Fig. 8.12b. Both kinds of stiffeners have predrilled
holes, but are attached to the pipe through holes which are made at the
site with a flame torch.

It is interesting to note that, for earlier applications, the transverse
stiffeners were considered necessary only for the control of the pipe
deformations during the backfilling operation. Consistent with the
belief that they were no longer useful after the completion of the backfill, these stiffeners were not galvanized like the rest of the shell, as
can be seen, for example, in Fig. 8.12a. The transverse stiffeners are
now believed to be necessary for the load-carrying capacity of the top
portion of the shell, and are galvanized accordingly (see Fig. 8.12b).

There are many soil-steel bridges with transverse stiffeners that are,
and have been, distress-free for many years. It is a tribute to the inno-
vation of engineers that these bridges were designed in the absence of
well-established analysis and design methods that take into account in
a rigorous manner the presence of the transverse stiffeners. Many of

 (a)

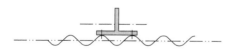 (b)

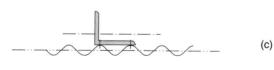 (c)

 (d)

Figure 8.11 Longitudinal sections through pipes showing various kinds of transverse stiffeners.

(a)

(b)

Figure 8.12 Photographs showing transverse stiffeners on the conduit wall: (*a*) corrugated plate stiffeners; (*b*) rolled steel beam stiffeners.

these bridges are known to have been designed by ad hoc and empirical approaches.

References 1 and 7 are recent contributions to the technical literature on the subject of transversely rib-stiffened shells of soil-steel bridges. Relevant details of these works are given subsequently in the hope that they will be found useful by designers in dealing with the design of soil-steel bridges with transverse stiffeners.

The aforementioned references show that, as illustrated in Fig. 8.13, the effective width of a stiffened plate decreases significantly with the increase in its curvature. This decrease in the effective width is so significant that, for the kinds of curvatures encountered in the conduit walls of soil-steel bridges, the effective width of the stiffened corrugated plate is approximately equal to the width of the stiffener itself. The reduction of the effective width with the increase of plate curvature is consistent with the behavior of curved beams. By using the classical theory of elasticity (e.g., Ref. 16), it can be readily demonstrated that the effective flange areas, and hence the flexural rigidity, of an I-beam that is curved in the plane of its web is smaller than that of a corresponding straight beam.

Reference 1 recommends that, for the purpose of analysis, the conduit wall stiffened with discrete transverse stiffeners be divided into stiffened and nonstiffened zones. As shown in Fig. 8.14, the stiffened zone is recommended to be limited to the width, W, of the stiffener. The remaining length of the conduit wall should be regarded as nonstiffened; the flexural rigidities per unit length of these zones are denoted as EI_2 and EI_1, respectively. For uniform spacing S between centers of the stiffeners, the equivalent flexural rigidity per unit length, EI_e is given by

$$EI_e = \frac{EI_2 W + EI_1 (S - W)}{S} \tag{8.5}$$

For obtaining estimates of the movement of the crown during backfilling, it is recommended that the pipe be analyzed by using the equivalent flexural rigidity. For dead- and live-load analyses of the completed structure, the equivalent flexural rigidity EI_e is first used in a similar fashion. Thereafter, the thrusts and moments thus obtained are multiplied by the respective values of γ (defined subsequently) to get the relevant values of these responses for stiffened and nonstiffened zones, respectively. The values of γ for the two zones are given by

$$\gamma = k \left\{ \frac{S}{R} \right\}^a \left\{ \frac{F}{f} \right\}^b \left\{ \frac{1 - \cos \dfrac{\alpha}{2}}{1 + \sin \dfrac{\alpha}{2}} \right\}^c \left\{ \frac{W\, EI_2}{S\, EI_e} \right\}^d \tag{8.6}$$

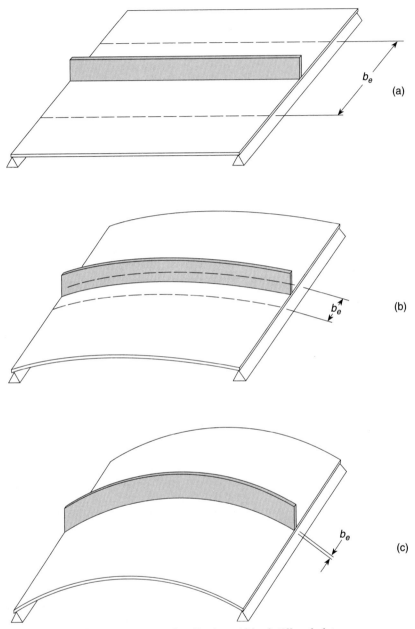

(a)

(b)

(c)

Figure 8.13 Effect of curvature on the effective width of stiffened plates.

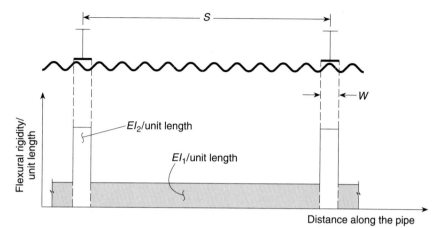

Figure 8.14 Assumed distribution of the flexural rigidities of stiffened and nonstiffened zones of the conduit wall.

where the coefficients k, a, b, c, and d are obtained for the two zones from Table 8.1, and the other notation is either defined earlier or given in the following list.

S = the center-to-center spacing of the stiffeners

R = the radius of curvature of the stiffened segment

F = the distance between the centroidal surfaces of the parent corrugated plate and the stiffener

f = the half-depth of the corrugation of the parent plate

α = the contained angle of the stiffened segment

The notation for some of the variables used in Eq. (8.6) is also illustrated in Fig. 8.15.

It is noted that, as explained in Ref. 1, Eq. (8.6) and the coefficients k, a, b, c, and d were developed by a process of curve-fitting through the

TABLE 8.1 Coefficients for Use in Conjunction with Equation (8.6)

Zone	Load effect	Coefficients				
		k	a	b	c	d
Non stiffened	Thrust	1.022	0.048	−0.059	0.023	−0.015
	Moment	1.083	0.027	−0.013	0.110	−0.052
Stiffened	Thrust	0.694	0.399	0.582	−0.806	−1.333
	Moment	0.487	0.584	−0.019	−1.207	−0.722

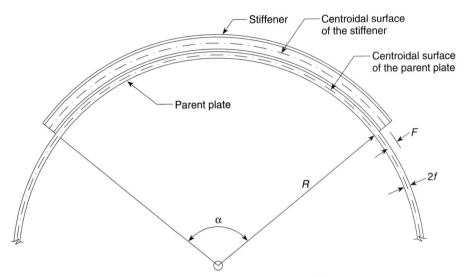

Figure 8.15 Notation used in the analysis of stiffened conduit walls.

results of a fairly large number of three-dimensional analyses of soil-steel bridges with transverse stiffeners. For the range of problems analyzed, Eq. (8.6) used in conjunction with the coefficients of Table 8.1 was found to have an error range of up to ±4 percent for nonstiffened zones and up to ±15 percent for stiffened zones. The preceding approximate method is given as an interim design aid until a more accurate method can be developed, possibly by first establishing dimensionless characterizing parameters for the interaction between stiffened and nonstiffened segments of the pipe.

8.2.2   Arch-Beam-Culvert (ABC) system

The Arch-Beam-Culvert (ABC) structural system was patented in Canada by C. W. Peterson (patent no. 1143170); it comprises a conventional soil-steel bridge in which the top segment of the pipe is made composite, through conventional shear connectors, with a reinforced concrete slab that projects transversely beyond the conduit. The anatomy of a typical ABC system incorporating a horizontally elliptical conduit is shown in Fig. 8.16. The system can, however, also be applied in conjunction with pipes of all shapes having either closed or open cross sections. The concrete slab carries a shallow fill on top of it. However, the concept is general enough to permit a slab whose top surface itself constitutes the riding surface of the roadway.

The ABC system is particularly useful when the depth of construction is limited and the conduit is required to have a relatively large

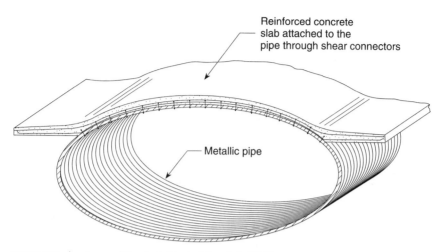

Reinforced concrete
slab attached to the
pipe through shear connectors

Metallic pipe

Figure 8.16 Anatomy of the Arch-Buried-Conduit (ABC) system.

span but small rise. A soil-steel bridge with such constraints and incorporating the ABC system is shown in Fig. 8.17. It is noted that the ABC structure is also known as Nova Span bridge (Ref. 4).

A relatively flat, reinforced concrete slab that is made composite with the top segment of the corrugated metal pipe can be designed as

Figure 8.17 A low-rise soil-steel bridge incorporating the ABC system.

the slab of a conventional slab bridge. For such a design, the reinforced slab can be assumed to be supported at its junction with the unstiffened slab. The pipe below the slab can be designed in the same manner as pipes without special features. It is recommended that further details of designing soil-steel bridges with the ABC system should be obtained from the literature of the company that holds the patent (Ref. 15) or from Ref. 4.

The adverse effect of long-term settlements of the foundation on soil-steel bridges with horizontally elliptically pipes has been discussed in Chap. 3. Such settlements can also be undesirable in structures with the ABC system and incorporating horizontally elliptical pipes. This is because the foundation on either side of such conduits is subjected to higher dead loads than the foundation under the conduit. Uneven settlements resulting from such nonuniform pressure on the foundation are likely to induce a gap between the projecting wings of the slab and the fill below. It is obvious that in these circumstances, the slab should be designed for the more pronounced cantilevering effects induced in its wings. It is possible to partly eliminate the differential settlements under the pipe and adjacent fill by providing a cushion of uncompacted soil under the pipe.

8.2.3 Concrete-Arch-Buried-Bridge (CABB) system

The quest for a solution to the problem of uneven foundation settlements discussed previously led to the development of the Concrete-Arch-Buried-Bridge (CABB) system. Similarly to the ABC system, the CABB system also relies on the strengthening of a segment of the pipe by attaching it to a layer, or shaped slab, of reinforced concrete. As shown in Fig. 8.18, the pipe of the CABB system incorporates a relatively flat bottom segment. The rest of the pipe above the bottom segment is made composite with the shaped slab through the kind of shear connectors that are customarily used in composite slab and girder construction. In its lower portions, the slab is thickened so as to act as a footing to the steel-concrete composite arch.

The distinctive feature of the CABB system is the joint between the steel-concrete composite arch and the bottom segment; this joint, through a system of slotted holes and pretensioned bolts with and without frictionless washers, permits a fairly large downward movement of the arch with respect to the bottom segment. A view of a typical joint used in the CABB system is shown in Fig. 8.19. This joint can permit a relative movement of about 90 mm (3.5 in). By controlling the tension in the bolts and the frequency of the frictionless washers, one can control the total maximum upward force that the bottom segment should sustain before the start of slippage in the joint.

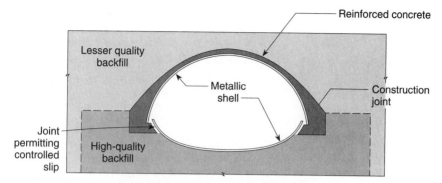

Figure 8.18 Anatomy of a Concrete-Arch-Buried-Bridge (CABB) system.

The CABB system is patented in Canada and the United States through patent numbers 1189332 and 4695187, respectively, by L. Mikhailovsky and J. G. Ramotar. Several structures have been built by using this system. One such structure built recently in the United States with a clear span of about 10.8 m (35.4 ft) is shown in Fig. 8.20. This structure was built by using 4.2-mm- (0.165 in) thick plates. It is noted that the thickness of the plate for application in the CABB system is governed mainly by requirements of welding the Nelson type of

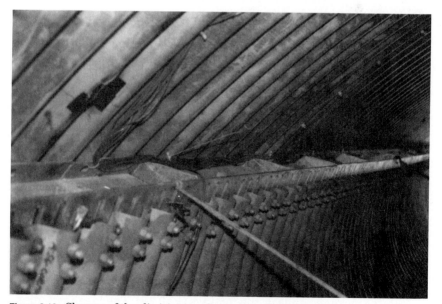

Figure 8.19 Close-up of the slip joint in a Concrete-Arch-Buried-Bridge.

Figure 8.20 Photograph of a Concrete-Arch-Buried-Bridge with a clear span of 10.8 m.

shear studs to it. While a thickness of 4 mm (0.157 in) was previously believed to be the minimum for such welding of 12.5-mm- (0.50 in) diameter studs, it is now known that even a thickness of 3 mm (0.118 in) is sufficient. Galvanization is usually applied to the corrugated plates after the installation of the shear connectors.

When the arch is relatively flat, the concrete slab above it can be cast without the help of a formwork. For steep arches, however, it is necessary to provide a formwork for their lower segments, as shown in Fig. 8.21. The reinforcement laid above the metallic shell for the arch can be seen in this figure, along with the formwork for the lower segment. After such formwork has been filled with concrete, the casting of the upper, flatter portions of the arch can proceed immediately. As shown in Fig. 8.18, the "footings" of the arch can be cast separately from the rest of the arch.

It will be appreciated that the wet concrete, until it has gained sufficient strength, is sustained entirely by the metallic shell which in itself is very flexible. Because of its high flexibility, it is important that the metallic shell not be subjected to large eccentric loads. Accordingly, the concrete around the shell should preferably be cast in a symmetrical manner.

The inventors of the CABB system prefer to analyze the arch as a frame by assuming that it is laterally supported only at its lower ends, i.e., at its thickened ends or the footings (Ref. 13). The idealization and

Figure 8.21 Formwork used for casting steep segments of the arch.

boundary conditions for such analysis are shown in Fig. 8.22. Consistent with the assumptions made in their analysis, the inventors recommend that, as shown in Fig. 8.18, the backfill providing horizontal restraint to the footings of the arch should be of high quality and well compacted; the rest of the backfill can be of a lesser quality. It is recalled that in the context of soil-steel bridges, a well-graded granular backfill is considered to be a high-quality backfill.

Neither the method of analysis nor the demarcation of the backfill on the basis of quality and degree of compaction are covered by the CABB patent. A designer should, therefore, feel free in designing the buried

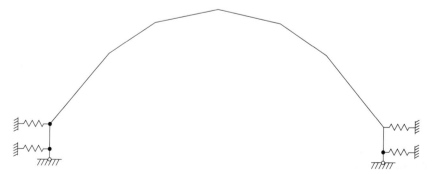

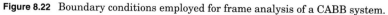

Figure 8.22 Boundary conditions employed for frame analysis of a CABB system.

concrete arch by any defensible method, and in specifying the backfill in conformity with the assumptions made in the calculations.

The limiting force which causes the joints to slip can be regarded as the thrust in the bottom segment. After the thrust is determined, the bottom segment can be designed by using the principles given in Chap. 4 for soil-steel bridges without special features.

8.2.4 Longitudinal stiffeners

There exist a large number of soil-steel bridges in which the pipes have been stiffened by longitudinal reinforced concrete beams located at each of their respective two shoulders; these beams are also known as *thrust beams,* presumably because the forces induced by the arch segment above cause them to thrust (i.e., react) against the backfill.

The reinforcement for a thrust beam is attached to the pipe as shown in Fig. 8.23. The installation of the formwork around the reinforcement and the casting of the beam usually take place after the backfill has been raised on both sides of the pipe to levels just below the location of the beam.

The proponents of the thrust beams claim that the vertical faces of these beams permit a better degree of compaction of the backfill in their vicinity. Other advantages claimed for the thrust beams are as follows:

Figure 8.23 Reinforcement cage for a longitudinal stiffening beam.

- By promoting a better distribution of live-load effects in the longitudinal direction, they reduce live-load thrust in the lower segment of the pipe.

- The thrust beams isolate the top segment of the pipe, thereby rationalizing the mathematical idealization that is commonly used in conjunction with its design.

While the authors have not themselves been able to quantify the benefits derived by the thrust beams, it is confirmed that a large number of soil-steel bridges with these special features are known to be performing satisfactorily. It is only a matter of conjecture to speculate that these structures might have performed just as satisfactorily without the thrust beams.

8.3   Special Features Which Reinforce the Backfill

As mentioned at several places in this book, it is imperative for the integrity of a soil-steel bridge that its backfill around the pipe continue to provide adequate support to the pipe during the lifetime of the bridge. Customarily, the adequate support to the pipe can be ensured by selecting a well-graded granular material for the backfill and compacting it to a dense and uniform medium. Realizing that such an ideal medium is sometimes difficult to attain, a few techniques have been developed to enhance the stiffness of the backfill without complete reliance on the compactive effort; these techniques are dealt with following.

8.3.1   Concreting under haunches

It has been discussed in Chaps. 3 and 10 that radial soil pressures in the haunches of pipe-arches are particularly high because of the relatively small radius of curvature of the conduit at these locations. In order for the soil to sustain these high vital pressures without yielding significantly, it is necessary that the backfill under the haunches be more densely compacted than anywhere else. Difficulty of access, however, makes it difficult to compact the backfill in these critical zones. Special techniques, such as water-jetting and rod-tamping, are available to compact the backfill under the haunches to required densities. However, even these techniques can be rendered ineffective in the absence of very strict and continual supervision.

Reference 3 has shown that, under the haunches of pipe-arches, the conventional compacted backfill can be replaced advantageously by low-strength and high-slump concrete. The extent of the concrete used in conjunction with one particular pipe-arch is shown in Fig. 8.24.

Before this technique was applied, many experts in the field of soil-steel structures were apprehensive of the floating up of the pipe during concreting and of the undesirable stresses that may be induced in the conduit wall at the junction of concrete and soil, i.e., at point A in Fig. 8.24.

The potential problem of the uplifting of the pipe during concreting was overcome in the case cited in Ref. 3 by pouring concrete in two layers and staggered longitudinal segments. Figure 8.25 shows the pipe under discussion with concrete poured in the first such segment.

Fears of damage to the conduit wall by the hard-point effect at the junction of concrete and soil, i.e., point A in Fig. 8.24, were also laid to rest by the performance of the structure, which does not show any sign of distress after six years of construction.

Low-strength and high-slump concrete has been proven to be a viable alternative to compacted backfill in restricted zones in pipe-arches; however, this material is too expensive for use as replacement for the entire engineered fill. The material discussed immediately following has all the desirable characteristics of high-slump concrete but is not as expensive and can be used to replace the entire engineered backfill, including the areas under the haunches of pipe-arches and other soil-steel bridges.

8.3.2 Controlled Low-Strength Material (CLSM)

Controlled Low-Strength Material (CLSM) is one of the many terms used for a flowable mortarlike mixture composed of granular soil, water, fly ash, and cement. This material is known by different names, such as flowable mortar and nonshrink grout, and is currently (1992)

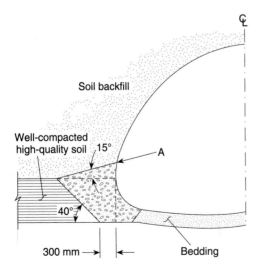

Figure 8.24 Part cross section of a pipe-arch showing the extent of concrete in one bridge.

Figure 8.25 Concrete poured around a haunch of a pipe-arch.

used mainly for backfilling road excavations. As pointed out in Ref. 5, several specifications exist for the proportions of CLSM; some of these proportions are given in Table 8.2.

It can be seen from Table 8.2 that about 72 to 88 percent of CLSM is composed of the soil which is required by many jurisdictions to pass 100 percent through 19-mm (0.75 in) sieve, and 0 to 10 percent through 0.074-mm (0.003 in) sieve. Cement proportions vary from 1.4 to 2.8 percent, and those of fly ash from 0 to about 11 percent. The water-cement ratio, ranging between 3 and 14, is extremely high compared to that used in concrete. When the water-cement ratio is so high, it has

TABLE 8.2 Proportions of CLSM by Various Authorities

		Weight in kg/m³ (Percentage by Weight) Specified by:				
Material	Ref. 5	American Concrete Pavement Association	Iowa Department of Highways	South Carolina Department of Transp.	Ohio Department of Transp.	City of North Platt
Soil	1543	1661	1543	1691	1727	1892
	(72.2)	(72.0)	(72.5)	(77.9)	(78.4)	(88.1)
Water	356	297	347	272	297	218
	(16.7)	(12.9)	(16.5)	(12.5)	(13.5)	(9.7)
Fly ash	178	250	178	178	148	0
	(8.3)	(10.8)	(8.2)	(8.2)	(6.7)	(0.0)
Cement	59	100	59	30	30	50
	(2.8)	(4.3)	(2.8)	(2.8)	(1.4)	(2.2)

virtually no effect on the strength of the mixture; the strength of CLSM is controlled instead by the cement and fly ash contents.

The CLSM sets within about 24 hours to nearly one-tenth of the 28-day strength. Depending upon the contents of cement and fly ash, the 28-day compressive strength of CLSM varies between 0.7 and 5.0 MPa (100 and 725 psi) It is interesting to note that in some CLSM applications, the emphasis is on keeping the strength sufficiently low to permit easy reexcavation.

Flowability.   The CLSM, having a very high slump of 160 to 200 mm (6 to 8 in), is so flowable that by gravity alone it can reach even those nooks and corners which cannot be accessed easily by compacting the backfill by conventional methods. To achieve greater flowability, the granular soil should have round, rather than angular, particles. In desert regions, the sand particles are usually well rounded, making them unsuitable for conventional backfill. This shortcoming of soils with rounded particles can be turned into an advantage by using them in CLSM, thereby achieving enhanced flowability.

Stiffness of backfill.   There are two distinct aspects of the stiffness of the backfill that are brought to bear on the load-carrying capacity of the metallic shell of a soil-steel bridge. One aspect relates to the arching action that controls the load effects in the shell, such as the thrust. The other aspect is concerned mainly with the stiffness of the radial support provided by the backfill to the metallic shell. As will be discussed later, the influence of the stiffness of the CLSM backfill is minimal on the load effects in the shell; however, its influence on the capacity of the metallic shell to sustain compressive axial loads is of paramount importance.

The measure of the stiffness of the backfill affecting the deformations and the load-carrying capacity of the pipe is a parameter called the modulus of soil reaction E'. As discussed in Chap. 7, the value of this parameter depends not only upon the engineering properties of the soil, but also upon the depth of embedment of the reference station. It can be readily demonstrated that a qualitative comparison of the E' values of different backfills having time-independent characteristics can be made simply by comparing the values of their respective moduli of elasticity.

It has been shown in Ref. 5 that CLSM with the mix proportions given in Table 8.2 (corresponding to the same reference) can develop a maximum compressive strength of 0.71 MPa (100 psi) and a mean modulus of elasticity of 6.5 MPa (950 psi) with a standard deviation of 3.3 MPa (480 psi). Despite their wide scatter, the values of the modulus of elasticity of CLSM can be represented realistically by the mean value. This is because the interaction of the shell and the backfill extends over a very wide area, because of which incipient buckling of a portion of the shell in the vicinity of a pocket of softer backfill is likely to immediately start internal re-arching, thereby averting buckling. The effect of such internal redistribution of the stiffness of the backfill support to the shell is equivalent to smearing the stiffnesses uniformly, in which case the mean stiffness can be used as the equivalent one.

A modulus of elasticity of 6.5 MPa (950 psi) is similar in magnitude to that of medium-grain soils compacted to between 85 and 95 percent standard Proctor density. It is thus concluded that the foregoing CLSM mix (i.e., that from Ref. 5, shown in Table 8.2) is similar in its stiffness characteristics to the well-compacted, good-quality backfill recommended for soil-steel bridges. Higher stiffnesses of CLSM can be achieved by rearranging their mix proportions.

Sequence of raising CLSM backfill. As discussed in Chap. 3, the backfill around the pipe performs two distinct roles. In one role, it is responsible for a very large portion of the load that the metallic shell is called upon to sustain, and in the other it provides the necessary support to the shell to enable it to sustain the induced load effects. In the case of conventional compacted backfill, a layer of loose soil simply adds gravity load on the shell; it is able to perform the latter, i.e., load-supporting, function only after it has been compacted. Since the conventional backfill is compacted in relatively thin layers, the load-sustenance aspect of the backfill does not lag far behind the load-including one. For such fills it is assumed quite justifiably that the backfill performs the two functions simultaneously.

The preceding assumption, however, is not valid for CLSM which loads the shell as soon as it is placed, but takes up to 24 hours to develop significant stiffness to be able to support the pipe. It is suggested that there are three critical stages governing the CLSM pouring sequence in soil-steel bridges having reentrant angles in the cross section of their conduits. The CLSM fill should be raised in at least three stages corresponding to these sequences, with at least 24 hours elapsing between the various stages. These three stages are discussed here with the help of Fig. 8.26.

For the first stage, the CLSM fill should be poured up to the spring line of the conduit. As shown in Fig. 8.26a, the CLSM fill can be contained transversely between compacted fills of lesser quality which can be raised to the crown level before the pouring of CLSM. As an alternative, the containment to the backfill can be provided by the sides of a trench that has been dug in existing ground. In both cases, however, the transverse dimension of the CLSM fill at the level of the spring line

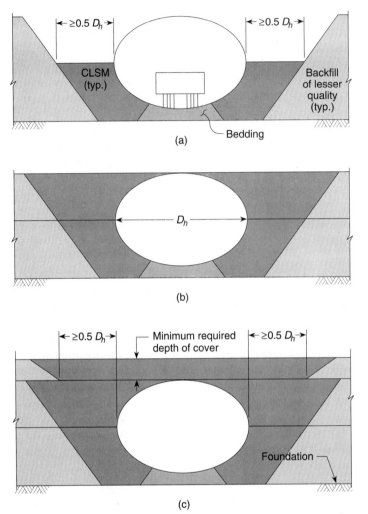

Figure 8.26 Three stages in pouring of CLSM requiring a minimum duration of 24 hours between them.

of the conduit should be at least one-half of the span of the conduit, as specified by many codes (e.g., Ref. 14).

The CLSM fill poured in the preceding stage lies under the pipe, thus exerting buoyancy forces on it. Until the CLSM in this lift is set, the pipe should be kept weighted down by placing a piece of heavy equipment (such as a truck) inside it, as shown in Fig. 8.26a. The weighting-down devices can be removed in about 24 hours after the completion of the first stage.

It is important that the CLSM fill be raised symmetrically on either side of the conduit. Failure to do so will result in undesirable rolling deformations of the pipe.

The preceding discussion about the first stage in the raising of the CLSM lift is clearly not relevant for arch structures without reentrant lower segments.

The second stage of raising the CLSM backfill should commence only after the backfill in the previous stage has gained sufficient strength; this takes place typically in about 24 hours. As shown in Fig. 8.26b, the CLSM backfill in the second stage should be raised up to the crown of the pipe. If the crown begins to rise too rapidly during this stage, it will be necessary to pour the material in thinner lifts and to give sufficient time intervals between the various lifts.

Until the backfill reaches the crown, the thrust induced in the pipe is not very high. However, the fill placed above the crown induces high thrust. It is for this reason that the CLSM backfill above the crown should be placed only after the lower backfill has gained sufficient strength to enable the metallic shell to sustain the additional thrusts.

All the fill above the crown can consist of CLSM. However, from considerations of economy, it might be preferable to limit the CLSM backfill to the minimum depth of cover discussed in Chap. 4. Above the cover of minimum thickness, the backfill may consist of soil of inferior quality, as it does in the case of conventional compacted backfills. As shown in Fig. 26c, the CLSM fill above the conduit can also be contained transversely by fill of lesser quality placed at a distance of about half the span from the outer boundaries of the conduit.

Dead-load analysis.   In soil-steel bridges with conventional compacted backfills, the dead-load effects in the metallic shell are obtained by taking account of the inherent stiffness of the backfill that contributes to the arching action discussed in Chap. 3. A lift of fluid CLSM induces load effects in the pipe long before it can affect the distribution of loads; consequently, it cannot participate in the arching action related to its own weight. It can be appreciated that the current methods of dead-load analysis, presented in Chap. 4, are developed with the assumption that the backfill participates in load distribution virtually as soon as it is placed; these methods are not suitable for structures with CLSM.

In the absence of a method developed rigorously for soil-steel bridges with CLSM, the dead-load thrust in the conduit wall of such structures can be obtained approximately by adapting Eq. (4.22) as follows, provided that the foundation of the structure is relatively unyielding:

$$T_D = 0.5W_l + 0.5(1.0 - 0.1C_s) A_f W_u \qquad (8.7)$$

in which, as shown in Fig. 8.27, W_l is the weight of the CLSM fill directly above the conduit, W_u is the weight of the rest of the fill directly above the conduit, and C_s and A_f are the terms already defined in Chap. 4.

Equation (8.7) is based on the obvious assumption that the weight of the wet CLSM directly above the conduit is sustained by the pipe without any transference of the load to the adjacent columns of the backfill. This assumption is valid only if the foundation does not have significant long-term settlements. If such settlements are anticipated, then the load transference can take place after the CLSM has set; in this case, it is safer to use Eq. (4.22) for the calculation of dead-load effects.

Hydraulic effects. As will be discussed in Chap. 9, soil-steel bridges carrying water through the conduit must have adequate inlet and outlet protection against damage by hydraulic forces. Part of this damage is related to the loss of fine particles caused by the water flowing through the fill behind the pipe. The CLSM is permeable, like the compacted granular backfill, but is less susceptible to the loss of fines because of its particles being held together by chemical, rather than frictional, bond. It is expected that the CLSM backfill is likely to be more resistant to hydraulic damage than the conventional backfill.

Economy. At a cursory glance, the CLSM might appear excessively expensive compared to the conventional compacted backfill. However, the perception of the large difference in the costs of the two materials is readily eliminated if it is realized that the cost of the compacted backfill includes the expenses associated with: (1) the effort to compact

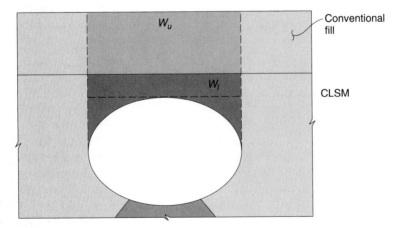

Figure 8.27 Definitions of W_u and W_l.

the fill in thin layers; (2) inspection and supervision of the compaction process; and (3) the testing of the soil density at different stages.

It has been demonstrated in Ref. 6 that in North America, the net cost of a CLSM backfill is only 4 percent more than the cost of well-compacted granular backfill. This cost comparison does not take into account the fact that a soil-steel bridge constructed with CLSM is less likely to be in distress than one constructed with the conventional backfill.

The authors believe that, especially in the context of lifetime costs, soil-steel bridges built with CLSM are likely to prove more economical than those built with compacted soil backfills.

8.3.3 Soil reinforcement by strips of steel

Reference 10 has proposed a novel concept of combining conventional soil-steel bridge design with the technique of reinforcing the soil with strips of steel. It is recalled that this technique of reinforcing the soil (known, for example, by the trade name of "reinforced earth") is used fairly extensively for retaining compacted earth fills in which steel strips are laid at different levels; these strips are tied to relatively thin facia panels.

The proposed concept is illustrated in Fig. 8.28, in which it can be seen that the steel strips are laid transversely to the axis of the pipe. The strips within the height of the pipe are connected to it, and those above run continuously up to a prescribed distance on either side of the conduit. Reference 10 describes tests on small-scale models incorporating the proposed concept. On the basis of these tests and extensive analytical studies, it is claimed that the reinforcing of the soil in the said manner will accomplish the following:

- reduce axial thrust and almost eliminate bending moments due to live loads in the conduit wall

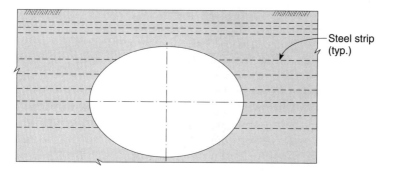

Figure 8.28 Cross section of a soil-steel bridge with soil reinforced by steel strips.

- generate multiload paths, thereby extending the active participation of the backfill in load sustenance of the pipe

- considerably reduce the movement of the backfill in the vicinity of the pipe, due to freeze-thaw cycles and temperature variations

- ensure that at ultimate loads, the structure will not fail in a catastrophic manner

- restrain the movement of the pipe during the backfilling operation

Notwithstanding the validity of these claims, the authors and many experts in the field of soil-steel structures believe that, as discussed subsequently, the last attribute mentioned may prove to be a disadvantage in certain cases, rather than an advantage.

In a conventional soil-steel bridge, the lateral pressures induced on the pipe during the early stages of backfilling cause the sides of the pipe to move inward, i.e., away from the soil, and the top to move upward. A reverse trend is set up during the later stages of backfilling, when the sides of the pipe are forced into the soil and the top moves downward. Experts in soil-steel bridges believe that the side segments of the pipe are able to mobilize the maximum support of the soil during the stage when these segments press into the soil. It is because of such belief that the unrestricted movement of the pipe during construction is considered to be an essential part of good construction practice, as discussed in Chap. 6. If the steel strips embedded in the soil are connected to the pipe, then the movement of the pipe during both the stages of backfilling will be restrained. It is feared that such restraint might not only prevent the development of full soil support to the pipe, but also create the hard-point effect at all the locations where the pipe is connected to the steel strips.

It is recalled that, as noted in Chap. 3, load effects in the conduit wall due to live loads are usually a very small fraction of the total load effects, except in structures with shallow covers. Because of this, the advantage of reducing live-load effects might not prove to be significant enough to encourage the selection of the system under consideration instead of a conventional one.

While there is some concern about the connection of the steel strips to the pipe, the provision of these strips above the conduit appears attractive because of their ability to avert a brittle failure and also because of their ability to reinforce the soil cover against a sliding-wedge-type failure. The latter consideration is responsible for code provisions relating to the minimum depth of cover. In the case of shallow covers, a CLSM fill used in conjunction with transverse steel strips might serve the same function as the relieving slab which is discussed in Sec. 8.1.3.

References

1. Abdel-Sayed, G., and Ekhande, S., "Rib-Stiffened Soil-Steel Structures," *Proceedings, 1st International Conference on Flexible Pipes,* Columbus, Ohio, 1991, pp. 7–15.
2. Bakht, B., "Live Load Response of a Soil-Steel Structure with a Relieving Slab," *Transportation Research Record 1008,* pp. 1–7, Transportation Research Board, Washington, D.C., 1985.
3. Bakht, B., and Agarwal, A. C., "On Distress in Pipe-Arches," *Canadian Journal of Civil Engineering,* vol. 15, no. 4, 1988, pp. 589–595.
4. Boully, G. K., and Plesiotis, S., "Investigation into Behaviour of Nova Span Bridge," *Proceedings of the AUSTROADS Bridge Conference,* Brisbane, Australia, 1991.
5. Brewer, W. E., "The Design and Construction of Culverts Using Controlled Low-Strength Material—Controlled Density Fill (CLSM-CDF) Backfill," *Structural Performance of Flexible Pipes,* Bolkema, Rotterdam, 1990.
6. Brewer, W. E., and Hurd, J., "Economic Considerations When Using Controlled Low-Strength Material (CLSM) as Backfill," paper no. 91-0309, *Transportation Research Record Annual Meeting,* Washington, D.C., 1991.
7. Ekhande, S., "Rib-stiffened Corrugated Soil-Steel Structures," Ph.D. thesis, Dep. of Civil Engineering, University of Windsor, Ontario Canada, 1986.
8. Gurfinkel, G., *Wood Engineering,* South Forest Products Association, New Orleans, Lousiana, 1973.
9. Katona, M. G., and Akl, A. Y., "Design and Analysis of Metal Culverts with Slotted Bolt Holes," *Department of Civil Engineering Report,* University of Notre Dame, 1984.
10. Kennedy, J. B., and Laba, J. T., "Suggested Improvements in Designing Soil-Steel Structures, *Transportation Research Record No. 1231,* Transportation Research Board, Washington, D.C., 1989.
11. Lefebvre, G., Laliberté, M. Lefebvre, L. M., Lafleur, J., and Fisher, C. L., "Measurement of Soil Arching above a Large-Diameter Flexible Culvert, *Canadian Journal of Civil Engineering,* vol. 13, no. 1, pp. 58–71.
12. Madsen, B., *Structural Behaviour of Timber,* Timber Engineering Limited, Vancouver, Canada, 1992.
13. Mikhailovsky, L., and Scanlon, A., "Construction and Load Testing of Concrete Arch Buried Bridge, *Proceedings, Annual Conference of the Canadian Society for Civil Engineering,* vol. 1, Quebec, 1992, pp. 273–282.
14. *Ontario Highway Bridge Design Code/(OHBDC),* 3d edition, Ministry of Transportation of Ontario, Downsview, Ontario, Canada, 1992.
15. Peterson, C. W., *ABC Structural System—Outline of the Concept and Its Applications,* W. Peterson Engineering Ltd., Edmonton, Alberta, Canada. 1985.
16. Timoshenko, S. P., and Goodier, J. N., *Theory of Elasticity,* McGraw-Hill, 1970.

9

Hydraulic Design

M. Saeed Choudhary

Ministry of Transportation of Ontario
Downsview, Ontario, Canada

Ron P. Parish

Bolter Parish Trimble Ltd.
Edmonton, Alberta, Canada

9.1 Introduction

In this book, a soil-steel bridge is referred to as a *culvert* when it conveys water through its conduit. The purpose of a culvert is to permit the passage of a predetermined extreme flood and associated ice and debris without jeopardizing the safety of the public on or around the bridge. The hydraulic design of a culvert entails the determination of its size, shape, and length with a view to providing a passage of water and to minimizing the following:

- adverse impact to flow alignment and other river processes
- harm to aquatic life and its habitat
- flooding or erosion of neighboring properties

The extreme flood condition for which a culvert is designed depends upon the hydrologic design criteria specified by the jurisdiction having authority over the bridge. Given the hydrologic design criteria and details of the terrain surrounding the structure, estimates of design flow corresponding to different return periods can be obtained by using well-established hydrological principles given in standard text books, e.g., Ref. 5.

9.2 Terminology

Recognizing that these may not be familiar to many structural engineers, some of the terms commonly employed in the hydraulic design of culverts and used in this chapter are defined here.

Afflux is the increase in the level of water upstream of the culvert, caused by an obstruction in the natural passage of water.

Constriction is the obstruction to the passage of water provided by a conduit which is narrower than the natural channel.

Critical depth is the depth of water at critical flow, defined immediately following.

Critical flow is the flow of water in the unstable transition state between streamlined and turbulent flows.

Critical slope is the slope at which critical flow occurs.

Culvert, as discussed earlier, is a soil-steel bridge which conveys water through its opening.

Design discharge is the discharge which a culvert is designed to accommodate without exceeding the constraints assumed in the design.

Discharge is the term used for quantifying the flow of water expressed as volume per second.

Fish passage is the portion of the conduit cross section which is designed to create favorable hydraulic conditions for migration and sustenance of aquatic life.

Flood is the term used loosely for relatively high flow in terms of either water discharge or its level.

Headwater is the term used for the pool of water upstream of the culvert.

Headwall is a wall at the end of a culvert, normally extending from the invert to either above or below the crown of the conduit and running either parallel to the roadway or perpendicular to the conduit axis.

Hydraulic jump is the abrupt rise in the depth of flow caused by change in the flow regime from supercritical to subcritical, which frequently occurs in the proximity of inlets and outlets of culverts.

Return period is the period in years during which a discharge may equal or exceed a given value.

Tailwater is the term used for the body of water downstream of the culvert.

Velocity head is the kinetic energy of flowing water expressed as its head, this being equal to $V^2/2g$, where V is the velocity and g is acceleration due to gravity.

9.3 Special Considerations

Before discussing the details of hydraulic design in a quantitative manner, it will be useful to discuss briefly the considerations, other than hydraulic, which must be brought to bear in the design of the conduit, or conduits, of a culvert. These special considerations are discussed subsequently under separate headings.

9.3.1 Size of conduit

The conduit should be large enough to permit the passage of the design discharge with a prespecified freeboard, the latter being the vertical distance between the lowest portion of the roadway and the crown of the conduit. Although the specific requirements may vary from site to site, it is desirable to examine the consequences of floods more severe than the design flood. For example, a culvert designed for a return period of 25 to 50 years, should also be evaluated with respect to the consequences of a 100-year flood.

9.3.2 Considerations for maintenance

Culverts should be designed in such a way that the exceedance of the design flood does not usually lead to damage of a kind which would require major effort in the restoration of the structure. It is understood, of course, that in certain conditions such a goal may be difficult to achieve.

9.3.3 Risk versus economy

The safety of the public is, indeed, paramount in the design of any structure. Without compromising such safety, it may be economically feasible to design a culvert with the knowledge that the road may become unusable for short periods due to submergence in extreme floods. Clearly, such designs can be permitted only for structures on low-volume roads for which alternative routes are available. When the road above a culvert is designed for occasional submergence, the savings in the cost of the structure should be carefully weighed against the anticipated risks and the cost of potential repairs.

9.3.4   Risk of blockage

Blockage of the passage of water by ice and debris in a culvert can cause a rise in the headwater elevation, which in turn may induce excessive flooding and overtopping, and high velocities and scour at the outlet. The conduit selected should be large enough to permit the passage of expected ice and debris. Measures for dealing with ice and debris are discussed in Sec. 9.5.

9.3.5   Impact on river processes

The conduit, or conduits, of a culvert should be so selected as to have a minimal impact on the natural river processes, which include erosion and sedimentation. For example, constriction of a wide natural channel by a small conduit might initiate backwater and the undesirable deposition of sediments at the upstream end of the culvert. Such a situation may also destabilize the balance in the transport of sediments, and thereby cause scouring at the downstream end.

9.3.6   Environmental considerations

The conduit selected for the culvert should cause minimal disruption to fish and other aquatic life. As will be discussed in Sec. 9.5, it may be found desirable in some cases to create fish passages by providing shallow baffles at the invert. It is noted that the provision of baffles reduces the hydraulic efficiency of the conduit, and that this reduced efficiency has to be compensated for by increasing the size of conduit.

9.4   Sizing of Conduits for Hydraulics

The hydraulic design of culverts entails the determination of the size of the conduit (or conduits, if there is more than one) from considerations of: (1) their hydraulic capacity, and (2) other factors discussed in Section 9.3. This section deals briefly with the former considerations.

When a constriction, such a culvert, is incorporated within a natural water course, the level of water rises upstream of the constriction, i.e., an afflux occurs. Permissible limits of the afflux for design discharges corresponding to different return periods are formulated by balancing risk with economy. In the absence of a set of universally accepted values, different jurisdictions have specified different limits of the afflux. For example, one jurisdiction may specify that the level of water upstream of the culvert must not rise above the crown in a 10-year flood but may rise up to the lowest elevation on the road in a 100-year flood. Another jurisdiction may specify that the water level at 25- and 50-year floods must remain 1.5 and 1.0 m, respectively, below the lowest elevation of the surface of the road over the bridge. A part of the hydraulic design of culverts entails the determination of the optimal number and size of the conduits so that such specified limits of the afflux are not exceeded.

9.4.1   Notation

The following notation, some of which is illustrated in Fig. 9.1, is employed in the hydraulic design of a conduit:

HW = depth of the headwater at the inlet to the conduit

TW = depth of the tailwater at the outlet of the conduit

S = slope of the conduit invert expressed as the ratio of vertical to horizontal projections of the slope

L = length of the conduit

H = difference in the elevations of headwater and tailwater

D = rise of the conduit

V = velocity of water at the reference section

g = acceleration due to gravity = 9.8 m/sec² (32.2 ft/sec²)

9.4.2   Flow controlled by inlet

When the gradient of the invert of a conduit is steeper than the critical slope, the water can flow through the conduit at a greater rate than that at which it can enter the inlet. In this case, the capacity of the conduit is unaffected by its length and the roughness of its inner surface. The flow in such a conduit is said to be controlled by the inlet.

Conduits with flow controlled by the inlet run partially full regardless of whether the inlet is submerged or not. An example of a conduit with submerged inlet and running with inlet control is shown in Fig. 9.2. In this figure, the hydraulic jump is shown to be occurring down-

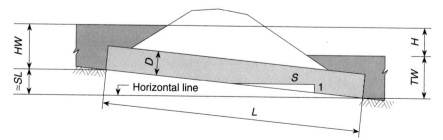

Figure 9.1 Illustration of notation used in hydraulic design.

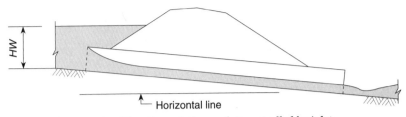

Figure 9.2 An example of flow through the conduit controlled by inlet.

stream of the pipe. There are also known cases in which the hydraulic jump occurs within the conduit in such a manner that the conduit runs full near the outlet.

The determination of headwater depth, *HW,* with inlet controlling the flow, is mathematically so complex that even experienced hydraulic engineers find it convenient to use the design nomographs of Ref. 11, which are available in several handbooks, e.g., Ref. 4. The effect of the level of discharge on the headwater depth of a circular conduit with a diameter, *D,* of 3.0 m (9.8 ft) can be seen in Fig. 9.3, in which *HW/D* is plotted against the level of discharge for three different inlet treatments. This figure shows that the relationship between *HW* and discharge is highly nonlinear and that end treatment can have a significant influence on the depth of headwater.

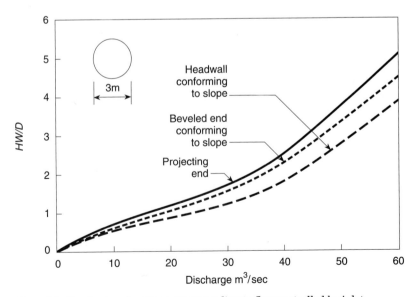

Figure 9.3 Headwater elevations corresponding to flow controlled by inlet.

Table 9.1 lists some of the inlet end treatments in descending order of their effectiveness in reducing the depth of headwater. As expected, the most efficient inlet treatment is that in which the pipe has a beveled end which is made integral with a sloping headwall. The least efficient end treatment is that in which the square end of the pipe projects beyond the fill. This, however, is also a treatment commonly used because it is economical.

From consideration of flow by inlet control, the optimum size of a conduit of a given shape is determined by an iterative process. Initially, a trial conduit is selected along with the details of the inlet treatment. Using the aforementioned nomographs, it is a relatively straightforward task to determine the value of HW for the specified discharge. If the value of HW thus obtained is higher or lower than the specified limit, the conduit size is respectively increased or decreased, and the whole process repeated until the optimum size is obtained. One can also change the inlet treatment to arrive at the optimum solution.

9.4.3   Flow controlled by outlet

When the gradient of the conduit invert is flatter than the critical slope, water can enter the conduit at a greater rate than the rate of flow through it. In this case, HW is said to be controlled by the outlet and is governed by several factors such as: (1) the shape and size of the cross section of the conduit, (2) the length of the conduit, (3) the gradient of the conduit invert, (4) the hydraulic roughness of the inside surface of the conduit, and (5) the depth of tailwater.

An example of flow through the conduit controlled by the outlet is illustrated in Fig. 9.4. In this example, the conduit is shown to be running full along its entire length. Under certain circumstances, the conduit can be partially empty along some of its length. Large culverts are usually designed for a condition in which the conduit or conduits run partially full to permit free passage of ice and debris.

With the help of Fig. 9.4, it can be readily demonstrated that HW is given by

$$HW = TW + H - SL \tag{9.1}$$

TABLE 9.1   End Treatments at Inlet in Descending Order of Their Effectiveness in Reducing HW

No.	Inlet end of conduit	k_e
1	Headwall with beveled end	0.2
2	Headwall with square end	0.5
3	Beveled end conforming to slope	0.7
4	Projecting from fill and without headwall	0.9

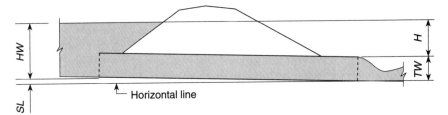

Figure 9.4   An example of flow through the conduit controlled by outlet.

where H is the loss of head. It is noted that Eq. (9.1) is valid only if the velocity head at the inlet is negligible as it usually is in the case of large culverts.

The tailwater depth, TW, is the depth of the water in the natural stream, i.e., the depth without constriction. When adequate information is not available to give a reliable estimate, TW is customarily assumed to be $D + 0.5\,Y_c$, in which D is the average of the span and rise of the conduit and Y_c is the critical depth.

The head H is a measure of the energy required to pass a given quantity of water through the conduit in the outlet mode of control. Standard textbooks on hydraulic design, e.g., Ref 5, have shown that H is composed of: (1) the velocity head in the conduit, (2) the entrance loss, and (3) the loss due to friction. These three components of H are represented, respectively, by the three terms within brackets on the right-hand side of the following equation:

$$H = \frac{V^2}{2g}\left\{1 + k_e + \frac{19.6 n^2 L}{R^{1.33}}\right\} \tag{9.2}$$

where V = the mean velocity of water in the conduit in m/sec

g = acceration due to gravity in m/sec^2

k_e = a coefficient which depends upon the configuration of the inlet; its values can be obtained from Table 9.1 for some end treatments commonly employed

n = the familiar Manning's roughness coefficient which depends upon the hydraulic roughness of the conduit surface with respect to its size; for 152×51 mm (6×2 in) corrugations, this coefficient varies between 0.028 and 0.033, with the lower and upper limits corresponding to conduits having diameters of 4.57 m (15 ft) and 1.52 m (5 ft), respectively; the value 0.028 can be used conservatively for larger conduits

R = hydraulic radius in meters being equal to A/P where A is the area of cross section of the body of water in the conduit and P is the wetted perimeter of the conduit

It is noted that the term in Eq. (9.2) which corresponds to the loss due to friction is applicable for metric units herein stipulated. For U.S. customary units, Eq. (9.2) takes the following form, with the units being in feet, as appropriate.

$$H = \frac{V^2}{2g} \left\{ 1 + k_e + \frac{29n^2L}{R^{1.33}} \right\}$$
(9.3)

The resistance to flow offered by the surface of the conduit wall is referred to as *friction,* probably for lack of a better term. Understandably, this friction is much larger in corrugated steel pipes than in pipes with smooth walls. It should be noted, however, that since no surface can be perfectly smooth, the smoothness or roughness of the inner surface of a conduit is only a relative measure of the irregularity of the surface with respect to the overall size of the conduit. A 51-mm- (2 in) deep corrugation might be regarded as providing a relatively rough surface for a conduit with diameter of 2.5 m (8.2 ft), in which case the depth of the irregularity is about one-hundredth of the mean conduit diameter. When the diameter is increased to, say, 10 m (32.8 ft), the irregularity in the surface is reduced to about one four-hundredths of the mean conduit diameter; in this case, the same surface can be seen as being smoother than that in the former case.

To obtain the optimum size of the conduit of a given shape from consideration of flow control by the outlet, the following information is needed:

- design discharge for the specified return period
- length, L, of the conduit
- slope, S, of the conduit invert
- detail of inlet treatment
- depth of tailwater

It is noted that the length of the conduit is governed by the depth and slopes of the soil cover above the conduit.

A trial size for the conduit is selected first, and the corresponding value of HW is determined by using Eq. (9.2) or (9.3), and Eq. (9.1) in conjunction with the foregoing data and appropriate values of k_e and n. The pipe size is fine-tuned iteratively so that the resulting value of HW is as close to the specified limit as practically possible.

9.4.4 Governing depth of headwater

Two values of HW are obtained for the optimal sizes of the conduit as discussed in the two prior subsections: one is for the flow controlled by inlet and the other for the flow controlled by outlet. For conduits with relatively small spans, the former case governs since it leads to higher values of *HW*. By contrast, control of flow by outlet governs in conduits of very large spans. For conduits of intermediate sizes, either inlet or outlet control may govern, depending upon the various factors.

9.5   Other Hydraulic Considerations

The capacity of a culvert to handle effectively the passage of water through its conduit or conduits is influenced by factors other than the size of the openings, the determination of which has been dealt with in Section 9.4. These other factors are discussed in this section under various headings.

9.5.1   Shape of conduit

While the shape of the cross section of the conduit is usually selected for reasons other than hydraulic, it is to be noted that, from hydraulic considerations, some shapes are preferred over others. For example, a cross section with greater width in the lower segments is the preferred one, because it can permit a given discharge with much lower depths than a circular cross section. For conduits with relatively large spans, a part-arch structure is preferable because of its providing a larger opening at lower levels. Such a structure is shown in Fig. 9.5. Similarly to part-arch structures, a pipe-arch has a wider opening at lower levels, as can be seen, for example, in Fig. 1.11. From a hydraulic standpoint, this latter structure is preferable to its circular and elliptical counterparts in the shorter-span range.

The effect of the shape of the conduit on ponding at the upstream end of the culvert can be seen in Fig. 9.6*a* and *b*. Figure 9.6*a* compares the levels of water in the natural stream and in a circular conduit of a given area of cross section. The same comparison is made in Fig. 9.6*b* with respect to a pipe-arch having the same area of cross section as the circular conduit. It can be seen that, in the latter case, the level of water in the conduit is lower than that in the former. It is for such reduction in the water level that pipe-arches are usually preferred over other conduit shapes of closed sections.

9.5.2   Multiple conduits

For a given situation, the choice between a single large conduit and multiple small conduits is not axiomatic. A single large opening has the advantage of permitting easily the passage of floating ice and debris such as driftwood. Multiple conduits, on the other hand, are

Figure 9.5 A large span part-arch conduit.

more efficient than single conduits in providing greater total cross-sectional area for the passage of water at lower depths. This aspect of multiple conduits can be readily appreciated with the help of Fig. 9.7, in which the total conduit areas below a reference level are plotted against the height of the reference levels for five different cases: (1) a single circular conduit, (2) a single rectangular conduit, (3) two circular conduits, (4) three circular conduits, and (5) a single semicircular conduit. For each case, the total area of the conduit or conduits is the same.

For a conduit combination to be efficient, the curve of the type shown in Chap. 8, Fig. 8.7 should be as close to the horizontal axis as possible.

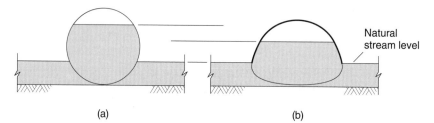

Figure 9.6 Effect of shape on ponding: (*a*) round pipe; (*b*) pipe-arch.

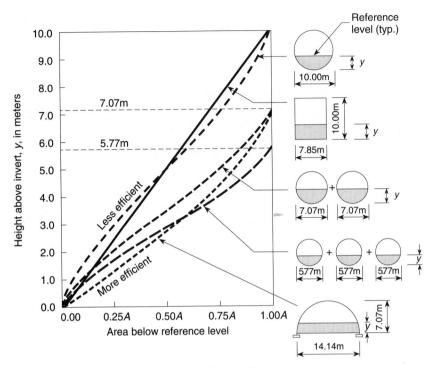

Figure 9.7 Efficiency of single and multiple conduits.

On this basis, it can be seen that the three-conduit combination is more efficient than the two-conduit one which, in turn, is more efficient than single conduits. As expected, the rectangular conduit is more efficient than the circular one. It is interesting to note that the single semicircular conduit is slightly more efficient than even the three-conduit combination. It is noted that multiple conduits are subjected to greater head loss than single conduits and, therefore, the curves presented in Fig. 8.7 may not be the only measure of hydraulic efficiency.

The choice between single and multiple conduits can be made only by experts using their judgment and experience in striking an economical balance between the advantages and disadvantages of the various alternatives.

9.5.3 Alignment

The alignment of the conduit axis is usually kept straight, regardless of the size and alignment of the natural stream. When a culvert is located on a bend in the stream or when the cross section of the channel is unusually wide, the channel is trained at both the inlet and outlet ends by means of earth berms. This is done to ensure that the water enters and exits the conduit with minimal disturbance and loss of energy. An example of such preferred alignment is shown in Fig. 9.8.

9.5.4 Velocity

Ideally, a culvert should be so designed as to have an insignificant effect on the natural velocity of water in the channel. In practice, however, the constriction provided by the conduit causes an increase in the velocities at the outlet. Experience has shown that unless the velocities at the outlet are below about 4.0 m/sec (13 ft/sec), there is a potential of damage by scouring, which must be accounted for through shaped outlets and energy dissipators.

It is noted that the exit velocity of the conduit is calculated by dividing the design discharge by that area of the conduit which lies below the surface of tailwater.

9.5.5 Invert level

The inverts of culverts are usually located about 0.6 to 1.0 m (2 to 3 ft) below the stream bed. One reason for this is to increase the cross-sectional area of flow at the outlet, and hence to reduce the exit velocity. Another reason is the sustenance of fish and other aquatic life by providing adequate water depths in dry periods and rest areas and passages for fish when the flow velocities are high.

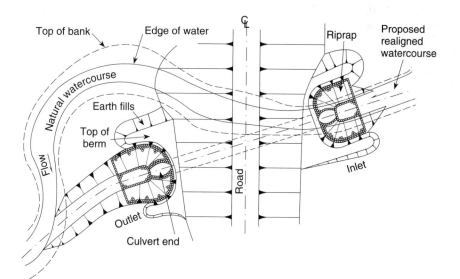

Figure 9.8 Plan of culvert layout showing preferred alignment and desirable end treatment.

9.5.6 Gradient of the invert

The gradient of the invert of a conduit should usually conform to the natural gradient of the watercourse. In the case of channels with exceptionally steep slopes, the culvert is sometimes laid at a flatter grade, mainly to dissipate energy and reduce the velocities at the outlet. It is interesting to note that, in conduits flowing with outlet control, the gradient of the conduit invert has very little influence on the depth of headwater.

9.5.7 Passage of ice and debris

As discussed earlier, the passage of ice and debris carried by flowing water must be considered in the hydraulic design of culverts. It is noted that debris carried by water does not consist only of driftwood. In mountainous regions, the flowing water can also carry stones of fairly large sizes through the conduit. Such a case can be seen in the photograph presented in Fig. 9.9. It is noted that when the debris consists of large stones, the main consideration (besides providing adequate room for ice) is that of protection of the conduit invert. Clearly, in such cases, an invert of bare, corrugated metal plate is undesirable.

In order to ensure that larger pieces of debris and floating ice do not block the conduit, two particular measures can be taken besides the

Figure 9.9 Large-size stones deposited in and around a culvert by flowing water.

selection of a single large conduit in preference to multiple small ones. These additional measures are as follows:

1. Training schemes to align appropriately the water which carries debris and ice, it being noted that a model study may be necessary for large projects

2. Provision of an area upstream of the culvert where debris and ice can be stored temporarily

9.5.8 End treatment

Both the inlet and outlet ends of the culvert and associated training works should be protected adequately. The protection at the culvert ends may consist of a full or partial headwall. A full headwall, along with integral wing walls, can be seen in Fig. 9.9, and a partial headwall in Chap. 3, Fig. 3.1.

The protection of the slopes of the earth berms, or guiding banks, can consist of rock riprap (which can be seen in Fig. 9.10) or other measures. At the upstream end, such protection should extend up to the design headwater elevation and, at the downstream end, the slope protection should be up to the level of the design tailwater elevation. It is

Figure 9.10 Photograph showing slope protection by rock riprap.

common practice to protect the slopes up to a length of one to three times the conduit span at the upstream end and up to a length of two to five times the conduit span at the downstream end. Particular attention should be paid to the protection of the outlets when potential exists for bed scour. The layout of a culvert presented in Fig. 9.8 also shows details of desirable end treatment.

9.5.9 Fish passages

Depending upon the presence of fish resources and the significance of the stream, it may be necessary to provide in the conduit resting areas and passages for fish; these may be provided by sinking the conduit invert below the stream bed, as discussed earlier. Additionally, it may be necessary in some cases to provide at the invert shallow baffles made of steel plates or other materials. In exceptional circumstances, fish ladders may be required.

9.6 Concluding Remarks

It is emphasized that this chapter has attempted to provide only an overview of the hydraulic aspects of culvert design; it does not contain enough information to enable one to perform the hydraulic design by considering all the relevant factors. Such designs should be performed by engineers who are well versed in both the art and science of the hydraulic design of culverts.

The information presented in this chapter has been derived from a number of sources, not all of which could be referenced conveniently. These references, which are listed below, will be found useful by those who wish to extend their skills in the hydraulic design of culverts.

References

1. Alberta Department of Transportation and Utilities, CULVFLOW, *Users and Programmes Manual,* Bridge Engineering Branch, Edmonton, Alberta, 1986.
2. AASHTO, *Model Drainage Manual,* chap. 9, "Culvert Task Force on Hydrology and Hydraulics," Washington, D.C., 1991.
3. AASHTO, *Technical Guidelines,* chapter on "Culvert Task Force on Hydrology and Hydraulics," Washington, D.C., 1991.
4. American Iron and Steel Institute, *Handbook of Steel Drainage and Highways Construction,* Washington, D.C., first Canadian edition, 1984.
5. Chow, V. T., *Handbook of Applied Hydrology,* McGraw-Hill, New York, N.Y., 1961.
6. Chow, V. T., *Open Channel Hydraulics,* McGraw-Hill, New York, N.Y., 1959.
7. Ministry of Transportation of Ontario, *Drainage Manual,* chapter D, "Culverts," Survey & Design Offices, Downsview, Ontario.
8. Morris, H. M., and Wiggart, J. M., *Applied Hydraulics in Engineering,* The Ronald Press Company, New York, N.Y., 1971.
9. Smith, C. D., *Hydraulic Structures,* University of Saskatchewan Press, 1984.
10. U.S. Army Corps. of Engineers, HEC-92, *Water Surface Profiles,* Hydrologic Engineering Center, Davis, California, 1992.

11. U.S. Bureau of Public Roads, *Capacity Charts for the Hydraulic Design of Highway Culverts,* Hydraulic Engineering Circular no. 10, Washington, D.C., 1965.

12. U.S. Bureau of Public Roads, *Hydraulic Charts for Section of Highway Culverts,* FHWA Hydraulic Engineering Circular no. 5, Washington, D.C., 1977.

13. U.S. Department of Transportation, *Bridge Waterways Analysis Model: Research Report* (WSPR), McLean, Virginia, 1988.

14. Wisneu, P. E., INTERHYMO/OTTHYMO 89, *Hydrologic Model for Stormwater Management and Flood Control,* Paul Wisneu & Associates Inc., Ottawa, Ontario.

Chapter 10

Distress, Monitoring, and Repairs

Baidar Bakht
John Maheu
Ministry of Transportation of Ontario
Downsview, Ontario, Canada

Buried corrugated steel pipes with diameters less than about 0.6 m (2 ft) are often constructed with a minimum of effort by placing the pipe in a roughly formed ditch and using the native soil as the loosely compacted backfill. This undesirable construction practice usually has no adverse effect on the performance of the small-diameter pipe, which is able to sustain the abuse because of having been designed mainly for handling rather than for strength. The construction practice employed for small-diameter pipes is sometimes also adopted for structures with larger spans without realizing that they do not have sufficient strength to withstand such deficiencies in the construction procedure.

Bad construction practice is the main reason for the incidence of structural distress being more prevalent in soil-steel bridges than in their conventional counterparts. The lack of understanding of the mechanics of behavior of these structures promotes bad construction practice.

Another reason for distress in soil-steel bridges is the lack of adequate inlet and outlet protection when the structure is called upon to convey water, i.e., when it is used as a culvert.

This chapter identifies the various forms of structural distress in soil-steel bridges, discusses the specific factors that may be responsible for them, and outlines the relevant means of repair and rehabilitation. It is noted that distress due to corrosion, which has already been covered in Chap. 2, is not discussed in this chapter.

The chapter also contains a section dealing with the monitoring of the cross-sectional shape of the conduit by photogrammetry. This technique is particularly useful in establishing quantitatively whether the deformations of the pipe are growing or have stabilized.

10.1   Forms of Distress

Structural distress in the metallic shell of a soil-steel bridge can manifest itself mainly in one or more of the following forms:

- Crimping of the conduit wall
- Bolt-hole tears
- Bearing failure at longitudinal seams
- Excessive deformation of the conduit cross section
- Lifting of the invert
- Lifting of the pipe ends
- Distortion of the pipe ends
- Collapse of the structure

These forms of distress are discussed in this chapter under separate headings. It is noted that washouts, in which the metallic shell of a soil-steel culvert can be floated away in severe floods, are not discussed.

10.1.1   Crimping of the conduit wall

Figure 10.1 shows a crimped segment of the conduit wall. Crimping can be regarded as a consequence of local buckling in which the metallic shell buckles into a large number of waves, each of relatively small length; it can occur in the compression zone of the wall section when the conduit wall undergoes large bending deformations. This kind of crimping usually takes place in conduit wall segments of relatively small radius of curvature; it indicates that the soil behind the segment is not dense enough to prevent excessive bending deformations.

Crimping can also occur in an entire conduit wall section subjected to excessive thrust while being supported by very well compacted backfill. Although the incidence of this kind of crimping is rare, it is known to have occurred in structures with circular conduits, which were constructed with good-quality, well-compacted backfill on relatively yielding foundation. It is postulated that the long-term foundation settlements of these structures induced negative arching, thus subjecting the conduit wall to greater and greater thrusts as time passed, until the thrust exceeded the buckling capacity of the conduit wall even though it was well supported.

Buckling of the entire conduit wall section into waves of small length has a redeeming feature. By reducing the axial rigidity and increasing axial deformations of the pipe, it induces positive arching, thus effectively reducing the axial thrust in the pipe. The result of this sequence

Figure 10.1 A crimped segment of the conduit wall.

is that, despite crimping, the pipe can be in stable condition provided, of course, that the time-dependent foundation settlements have ceased.

If the only sign of distress in a soil-steel bridge is crimping limited to a few segments, then in most cases one need not be too concerned about the structural integrity of the structure.

10.1.2 Bolt-hole tears

Some results of a survey of distress in soil-steel bridges, reported in Ref. 2, are reproduced in Fig. 10.2a, b. Figure 10.2a shows that pipe-arches are more prone to distress than the other kinds of soil-steel bridges. It is noted that this observation is not the result of an unusually large number of pipe-arches included in the survey. Figure 10.2b further shows that in pipe-arches, the most common form of distress is that which is identified as "bolt-hole tears."

Bolt-hole tears is another name for cracks in the bolt holes in longitudinal seams. As shown in Fig. 10.3, these cracks are usually horizontal. Since the conduit wall is always subjected to compressive forces, the bolt-hole tears usually do not extend over the entire section of the wall. There are, however, some known cases in which the bolt-hole tears, by eventually extending over the entire section, have completely severed large segments of the conduit wall.

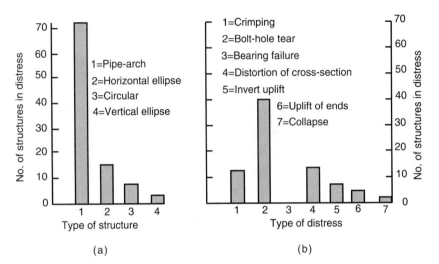

Figure 10.2 Results of a survey of distress in soil-steel bridges: (a) distribution of the types of structure in distress; (b) distribution of the types of distress.

Reference 2 reports that, to study the mechanics of the development of bolt-hole tears, two straight corrugated plates were lap-jointed and tested for bending according to the scheme of Fig. 10.4. The plates were joined by a double row of 20-mm- (¾ in) diameter bolts, with the same pattern as is used in typical soil-steel bridges. As shown in Fig. 10.4,

Figure 10.3 Close-up of bolt-hole tears.

there were 11 bolts across the joint, the tension in them varying linearly from 136 N·m (1200 lb·in) in the bolt at one end to 373 N·m (3300 lb·in) in the bolt at the other end.

The load in the bending test was gradually increased until cracks appeared in the hole carrying the bolt with the highest tension, which is located in the row of bolts identified as CC in Fig. 10.4. The general shape of the cracks was the same as the bolt-hole tears observed in the field. Further increase in the load resulted in the development of cracks in the next hole in the same row, and then in the next one, and so on until all holes in this row had cracked. Bolt holes in the other row, i.e., BB, had no cracks.

Upon removing the bolts after the test, it was discovered that in each case a crack initiated from a sharp dent made in the plate by either a bolt head or a nut. It is noted that both the bolt head and one side of the nut have chamfers as shown in Fig. 10.5. These chamfers, which are supposed to eliminate the possibility of making sharp dents in the plate, clearly do not achieve this objective.

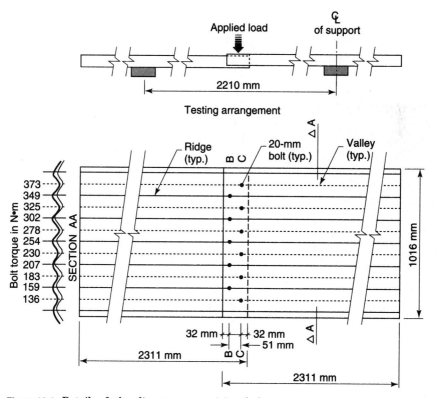

Figure 10.4 Details of a bending test on two jointed plates.

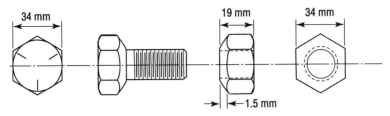

Figure 10.5 Details of a typical bolt and nut used in joints of the conduit wall.

The very simple test convincingly demonstrated that the bolt-hole tears are caused by excessive bending of the plates, and that the tendency to develop the cracks is directly related to the tension in bolts.

Figure 10.6 shows the exaggerated deformation of two jointed curved plates in bending. It can be visualized readily that the bending of plates causes them to part at the joint as shown in this figure. This parting increases the tension in the bolts. The increase in the bolt tension, however, is larger for bolts in one row than for the bolts in the other. For example, in the case shown in Fig. 10.6, the increase in tension for bolts along section CC will be higher than that for bolts along section BB. If the bolts with higher tension are placed on those faces of the corrugation which are themselves in flexural tension, the likelihood of the sharp dents in the plate propagating into cracks is increased. By corollary, this likelihood can be reduced by placing the higher-tension bolts on those faces of the corrugations which are in flexural compression.

In light of the preceding discussion, the bolting arrangements shown in Fig. 10.7*b*, which are often employed in practice, are found to be undesirable. The desirable bolting arrangement is that shown in Fig. 10.7*a*.

Alberta Transportation Department in Canada has conceived a clever and foolproof guideline to ensure that a desirable bolting

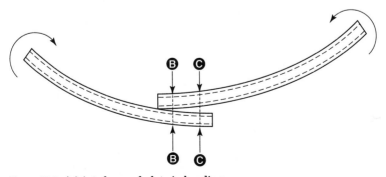

Figure 10.6 A jointed curved plate in bending.

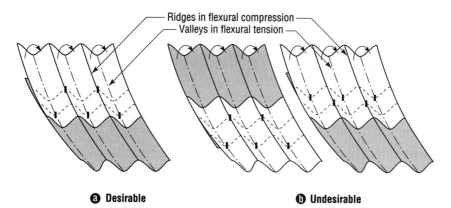

Figure 10.7 Bolting arrangements: (*a*) desirable; (*b*) undesirable.

arrangement is adopted in all cases. According to this guideline, the bolts in the row close to the visible edge of the mating plates should be placed in the valleys, and those in the other row, on the ridges. It is interesting to note that the guideline remains valid regardless of whether the joint is viewed from inside the conduit or outside. The bolting arrangement also remains unaffected by the sign of the moments. Further discussion on bolting arrangements is provided in App. B.

As discussed earlier, bolt-hole tears are most common in pipe-arches at the longitudinal seam between the top and side segments of the conduit wall. However, they are also found, albeit not so frequently, in other soil-steel bridges. It is noted that the bolt-hole tears are not always the result of excessive deformation of the conduit wall of the completed structure; they can also be formed during the pipe assembly when poorly matching plates are forced to fit at the longitudinal seams.

It is emphasized that even the use of the desirable bolting arrangement just discussed is not an effective means of avoiding distress if the backfill behind the longitudinal seam under consideration is poorly compacted.

10.1.3   Bearing failure at longitudinal seams

Bearing failure at longitudinal seams can take place due to the yielding of the conduit wall directly under the bolts. Clearly, this form of failure takes place under excessive conduit wall thrust and in conditions which prevent excessive bending deformations.

While the bearing failure of longitudinal seams has been observed in laboratory testing of the strength of bolted joints, it is extremely rare to find it in practice.

10.1.4   Excessive deformation of conduit cross section

Excessive deformation of the cross section of the conduit, although not the most common, is the most visible form of distress in soil-steel bridges; it causes considerable alarm regarding the safety of the structure. An example of excessive pipe deformation is shown in Fig. 10.8. Excessive deformations of the conduit wall are caused by the inability of the backfill to restrain its movement. This inability usually results from one or more of the following factors:

- Poorly compacted backfill
- Backfill containing large quantities of clay or organic matter
- Well-compacted and good-quality backfill not extending far enough on either side of the conduit

Excessive pipe deformations do not always develop after the structure has been built. Because of its flexibility, the pipe can deform excessively during the initial stages of the backfilling operation. If such deformation is not prevented or corrected during construction, the structure is built with deformed pipe. Pipe deformations locked in during construction may not be detrimental to the structural integrity of the structure, especially if they have stabilized. Pipe deformation occurring after the completion of the structure may, on the other hand, be a warning signal for the imminent collapse of the structure.

Figure 10.8 Photograph showing excessive deformation of the pipe.

If a record is not kept of the exact shape of the conduit of the as-built structure, it becomes difficult at a later stage to distinguish between the pipe deformations locked in during construction and those which are induced later due to inadequacies of the backfill. There are known cases in which a soil-steel bridge is declared unsafe, quite wrongly, several years after its construction because conduit wall deformations which were locked in during construction were considered to have occurred later.

It is very important that a record of the as-built conduit shape be kept so that it can be ascertained later whether the observed deformation occurred recently or has been there since the construction of the structure. Photogrammetry, discussed in Sec. 10.3, will be found useful in recording the shape of the conduit of the as-built structure.

When the records of the as-built structure are not available, it is important to record the changes in the conduit shape at regular intervals after the deformations were first noticed. If the deformations are not significant and have not undergone significant changes, then it is likely that the structure is safe. Local deformations of the conduit section tend to have a self-redeeming quality. As illustrated in Fig. 10.9, these local deformations may cause the soil to arch around them, thus relieving the pressure which led to them.

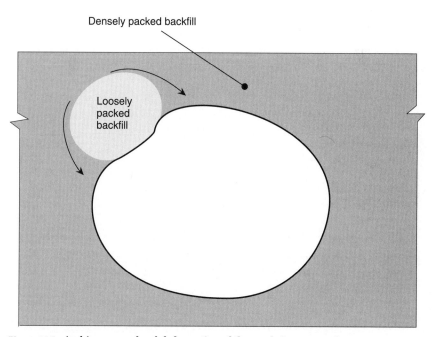

Figure 10.9 Arching over a local deformation of the conduit cross section.

It should be noted that in many cases the collapse of a soil-steel bridge is preceded by noticeable and rapidly deteriorating distortion of the conduit cross section. Therefore, if the conduit distortions are found to be worsening with time, it is important either to close the bridge or take precautions to avoid the sudden collapse of the structure.

10.1.5 Lifting of the invert

One of the forms of distress usually encountered in pipe-arches is uplift of the invert. The main cause of the occurrence of this form of distress can be explained readily with the help of Fig. 10.10.

As shown in Fig. 10.10, the vertical pressures at the horizontal plane just below the pipe are far from uniform. The pressure under the invert is well below the free-field pressure that exists away from the pipe; however, the vertical pressures directly under the haunches are significantly higher. The consequence of this situation is that, if the foundation is relatively yielding, the pipe settles more under the haunches than under the invert, thus leading to the uplift of the invert as illustrated in Fig. 10.11.

The uplift of the invert can in certain cases cause the longitudinal seam in the invert to lose its ability to transmit the axial force from one mating plate to another. Some structures are still standing even after the uplift of the invert has severed the longitudinal seam in that area. The reason why the structure can still stand in this condition is the development of high radial pressures in the haunch area, as illustrated in Fig. 10.12. The structure shown con-

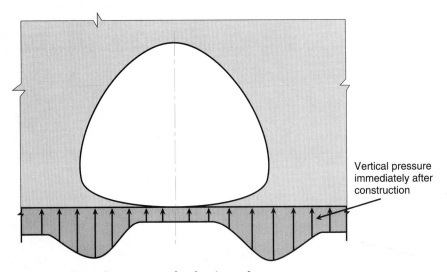

Vertical pressure immediately after construction

Figure 10.10 Vertical pressures under the pipe-arch.

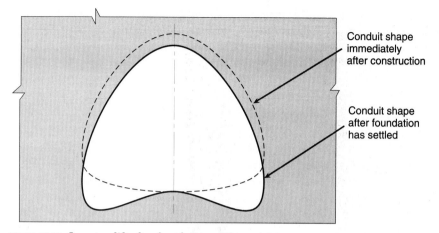

Figure 10.11   Invert uplift after foundation settlement.

ceptually in this figure acts like an arch structure, the difference being that the lower supports of its conduit walls are tenuous and cannot be relied upon as the footings of an arch structure. In fact, if the cause of the invert uplift is the time-dependent deformations of the foundations, then it is virtually certain that the soil under the haunch areas will eventually yield, causing either collapse or excessive deformations of the pipe.

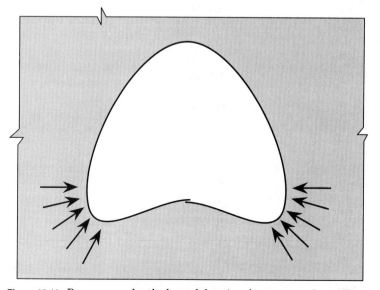

Figure 10.12   Pressures under the haunch keeping the structure from collapsing.

To avoid invert uplift, it is essential that the pipe-arch be either built on relatively firm foundation or that the foundation under the haunches be strengthened. One possible way of strengthening the foundation for pipe-arches is shown in Chap. 4, Fig. 4.8.

10.1.6 Lifting of the pipe ends

Another form of distress commonly found in soil-steel bridges occurs in structures in which water flows through the conduit. As illustrated in Fig. 10.13, this kind of distress is caused by a combination of uneven settlement of the pipe foundation along its length and buoyancy effects. The uneven foundation settlements occur because the pipe on its middle length carries more soil load than on its ends where the soil embankment is sloped. The buoyancy forces act only when the water is allowed to seep below the invert. The erosion of the soil cover, as shown in Fig. 10.13, worsens the situation by removing the weight of the soil above the pipe end, which previously helped to keep the pipe from uplifting.

As discussed in Chap. 3, Sec. 3.2, because of the passive resistance provided by the soil in the corrugations, the fully embedded corrugated pipe is quite stiff with respect to longitudinal cantilever bending which causes the pipe ends to uplift. A segment of such an embedded pipe is shown in Fig. 10.14a. Figure 10.14b, c show the ends of unbeveled and beveled pipes, with little or no protection against the flow of water under the invert. As shown, the seepage of water causes the invert of the pipe to separate from the underlying soil, thereby reducing the flexural rigidity of the pipe against the cantilevering bending moments induced by the buoyancy forces. Since the pipe ends do not carry any soil dead weight above the crown, the uplifting of the ends becomes even easier.

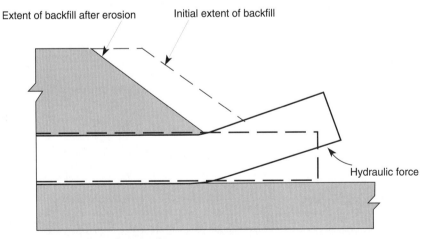

Figure 10.13 Uplifting of pipe ends.

The uplifting of the ends takes place only in those hydraulic structures in which the inlet and outlet ends of the pipe are not protected adequately. Figure 10.15 shows an example of a soil-steel bridge without such protection. The provision of cutoff walls which are installed under the culvert, as illustrated in Fig. 10.16, prevents the flow of water under the pipe, and thereby prevents not only the migration of fines from the bedding, but also reduces the buoyancy forces acting on the pipe. Figure 10.16 shows the ideal inlet treatment which includes the headwall, the cutoff wall, splayed wing walls, and the apron, all of which, however, are not necessary for all situations.

10.1.7   Distortion of beveled ends

To prevent the ends of a pipe from being unsightly, they are beveled, as shown schematically in Fig. 10.17. A view of the beveled ends of two pipe-arches is shown in Chap. 1, Fig. 1.22. At its beveled ends, a pipe is particularly vulnerable to damage by horizontal pressures. A complete pipe, because of having a closed section as shown in Fig. 10.18a, can sustain much higher intensities of the lateral pressure than an incomplete ring, shown in Fig. 10.18b. Lacking a closed section, the beveled ends of a pipe are prone to damage by heavy equipment pieces falling on them or by lateral earth pressures.

The usual method of supporting the bevel ends is to tie them through steel cables to wooden planks which are embedded in the backfill as deadmen (shown in Fig. 10.19).

10.1.8   Collapse

Collapse is the most dramatic manifestation of distress in soil-steel bridges. In some cases, such as the one shown in Fig. 10.20, only a partial length of the pipe collapses; in others, the whole pipe can collapse.

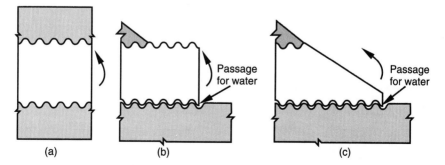

Figure 10.14 Resistance to uplifting by moment arm: (a) pipe in the middle segment with high resistance; (b) unbeveled end with low resistance; (c) beveled end with low resistance.

Figure 10.15 An example of a structure without a concrete head wall.

It is matter of great concern that the incidence of collapse is more common in soil-steel bridges than in other bridges. What is also of concern is the fact that many failures of soil-steel bridges could have been avoided by constructing these structures carefully, using good-quality backfill and employing good construction practice.

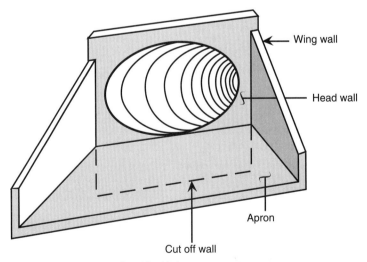

Figure 10.16 Anatomy of an ideal inlet treatment.

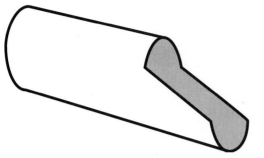

Figure 10.17 A pipe with a beveled end.

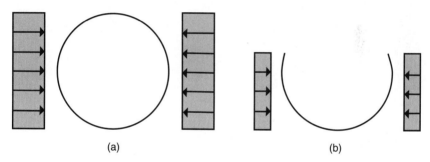

(a)   (b)

Figure 10.18 Distinction between a closed and open ring: (*a*) closed ring; (*b*) open ring.

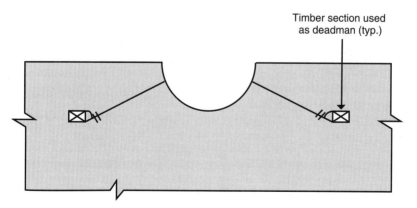

Timber section used
as deadman (typ.)

Figure 10.19 Use of deadman to support beveled ends.

Figure 10.20 Photograph showing the partial collapse of a soil-steel bridge.

Some of the factors that lead to the collapse of a soil-steel bridge are as follows:

- Use of poor quality soil, containing large quantities of clay and organic matter, in the backfill
- Compaction of the backfill in very thick layers
- Compaction of the backfill in very cold weather, when there are ice lenses in the soil which give rise to the false impression of an adequate degree of compaction
- Lack of compaction in areas where the interface radial pressures between the soil and conduit wall are particularly high
- Construction of the structure on very flexible foundation without strengthening it as required
- Providing skewed bevel ends to the pipe without adequate protection in the form of strong head walls made integral with the conduit wall
- Lack of inlet and outlet protection when the structure carries water and is expected to be subjected to sudden and severe floods

It is emphasized that properly designed and constructed soil-steel bridges are virtually maintenance free and show no sign of distress despite being in service for long periods. The incidence of distress, including collapse, in soil-steel bridges can be virtually eliminated by ensuring that these structures are designed and built properly under competent supervision.

10.2   Remedial Measures

The various measures that have been used in the past to deal with the problem of distress in soil-steel bridges are presented in this section, along with those techniques which are still in an experimental or conceptual stage. While most of the methods presented fall into the category of repair, others can be regarded as means of ensuring that the structure does not collapse suddenly with catastrophic consequences.

10.2.1   Temporary props

One of the most effective and expedient measures to ensure that excessive deformations of the pipe do not degenerate into sudden collapse is the provision of temporary struts or props in the conduit. These props can be timber columns of about 200 × 200 mm (8 × 8 in) cross section, or steel struts of hollow circular section of the kind used in construction formwork. As illustrated in Fig. 10.21, the props are located in the conduit under the crown and are provided with longitudinal sills above and under them. The sills, which run along the conduit length, are of timber. When the sides of the conduit cross section are also excessively deformed, the vertical props are supplemented with horizontal supports, as also shown in Fig. 10.21.

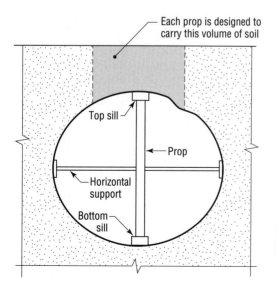

Figure 10.21 Cross section of a soil-steel bridge showing temporary props.

The main advantage of vertical props is that they can prevent a catastrophic failure of the structure; the main disadvantage is that they constrict the conduit. This disadvantage can be particularly significant for culverts.

The props should be designed to carry, with an adequate margin of safety, the weight of that volume of soil which is statically apportioned to them by assuming the propped cross section will act as a two-span continuous beam. The cross section of the volume of soil that a temporary prop should be designed to sustain is shown conceptually in Fig. 10.21.

The props are usually spaced at 1.0 to 1.5 meters. The butt joints of the top and bottom sills should be staggered so that they do not occur at the same location along the pipe. The sills should be long enough to contain at least two props. Figure 10.22 shows a soil-steel bridge supported by a crude framework of timber. Steel props are also used sometimes. Screws for adjusting the lengths of these props are very effective in ensuring that the contact between the supports and the pipe is not loose.

10.2.2 Partial concreting inside conduit

As discussed earlier, the most common form of distress in soil-steel bridges, especially pipe-arches, is the occurrence of bolt-hole tears in

Figure 10.22 Photograph showing a soil-steel bridge in distress supported by steel props.

longitudinal seams close to the invert. This form of distress indicates the presence of relatively loose fill behind the haunch areas. Accordingly, the most appropriate means of repair appears to be one which includes consolidation of the loose fill by some technique.

Unfortunately, a technique for consolidating the loose fill behind the haunches has not yet been proven by field application. In the absence of such a technique, the next-best repair method appears to be one in which the conduit wall is not only reinforced to transmit shear at the section containing bolt-hole tears, but is also made flexurally very stiff at haunches. This can be achieved by partial concreting of the inside of the conduit at the haunches, as illustrated in Fig. 10.23a.

Effective contact between the conduit wall and concrete can be provided through shear studs which are machine-welded to the pipe after grinding off locally the zinc coating of the galvanized plate. As will be discussed later in the section and shown in Fig. 10.28, shear studs can also be installed on the outside of the conduit.

It can be appreciated that, at haunches, the conduit wall has a tendency to bend in such a way that tension occurs towards the outside of the conduit. Except in the vicinity of the bolt-hole tears, the conduit wall itself can sustain the tensile forces. However, in the vicinity of bolt-hole tears, a reinforcing mesh should be provided close to the conduit wall, as illustrated in Fig. 10.23a. It should be noted that this reinforcing mesh should not only sustain the tensile forces resulting from subsequent excessive bending of the wall, but be capable of transmitting the shear forces acting on the wall at the location of bolt-hole tears.

The concrete can be cast in two lifts. For the first lift, it can be poured up to the horizontal construction joint shown in Fig. 10.23a. The second lift, requiring nearly vertical shuttering, can be cast later. Alternatively, and more effectively, the concrete in the haunch areas can be applied through shotcreting. Figure 10.23b shows the cross section of a pipe-arch repaired in 1992 in Ontario through shotcreting.

Fig. 10.23c shows another method of partial concreting in which the concrete at the two haunches is extended to cover the invert as well. This form of repair is particularly useful for those pipe-arches which suffer not only from bolt-hole tears, but also from minor invert uplift. As shown in Fig. 10.23b, the concrete in the invert area should be provided with a reinforcement mesh close to the top surface of the concrete, to sustain tension arising from further uplift of the invert. It can be appreciated that this repair technique may not be suitable if the invert uplift is excessive.

The significant disadvantage of the foregoing repair techniques is the reduction of the conduit size; this may be particularly undesirable in the case of culverts. It is noted, however, that this reduction is much

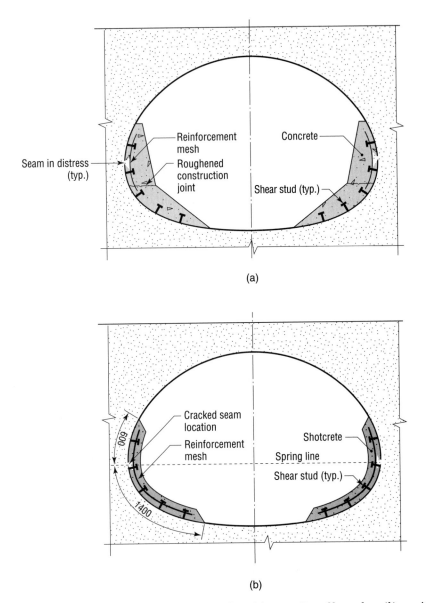

Figure 10.23 Repair techniques for pipe-arches: (*a*) concreting of haunches; (*b*) repair by shotcreting of haunches; (*c*) concreting of haunches and invert.

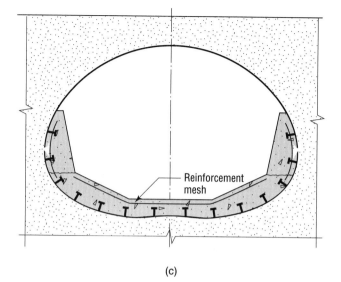

(c)

Figure 10.23 (*Continued*)

smaller if the methods of Fig. 10.23*a* or *b* are used, in which the concreting is limited to haunches.

10.2.3 Internal grouting

A very expensive, but effective, method of repairing a soil-steel bridge with excessive deformations is to place a smaller pipe within the existing one and fill the space between the two pipes with concrete grout. This method of repair, which has been successfully employed in several jurisdictions (e.g., see Ref. 4), is particularly useful when closing the structure for repairs for extended periods is not feasible. It is also an economical solution when the depth of cover over the conduit is large, making the removal of fill for replacement of the damaged segments of the pipe an even more expensive proposition.

Repair by placing a smaller-size pipe inside the existing conduit and internal grouting is feasible for structures having spans up to about 3 m (10 ft). For this method, a concrete floor or a pair of rails is first installed at the bottom of the existing pipe. The new pipe, which has been erected outside the conduit, is then dragged through the existing pipe on the concrete floor or rails already installed. The new pipe is just large enough to fit through the existing pipe with a prescribed gap of 150 to 200 mm (6 to 8 in). The gap between the two pipes is maintained through strategically located spacers.

The concrete grout is injected in the gap between the two pipes through two holes in the pipe, located in the shoulder areas, as shown in Fig. 10.24. After the grout has set, the repaired structure usually becomes much stronger than the original one, and remains virtually free of distress.

For structures having spans greater than about 3 m (10 ft), the preceding procedure may not be a practical one because of the large size of the internal pipe which must be drawn through the existing one. In such a case, the internal metallic shell can be constructed from steel tunnel liners.

The bottom plates of the liner are first connected to the invert of the existing pipe by means of anchor bolts. A gap of 150 to 200 mm (6 to 8 in) is maintained between the existing pipe and the liner plates through steel chairs welded to the invert.

After the gap between the bottom liner plates and existing invert is grouted, the liner plates are added symmetrically on both sides of the cross section, until the ring in completed by welding the top segments. The liners are provided with holes in the shoulder areas for injection of the grout.

Repair through liner plates permits the removal of locally deformed plates and their replacement by new ones. However, when this operation is undertaken, it is advisable to install complete rings of the liner plates on both sides of the plate to be replaced. Only after the grout between these complete liner rings and the existing pipe has set should an attempt be made to remove the damaged plate. After the damaged plate is removed, a portion of the soil will fall naturally, leaving a cav-

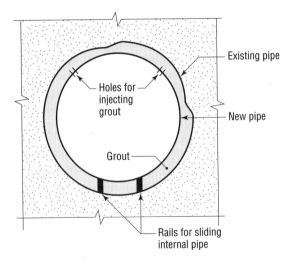

Figure 10.24 Repair by internal grouting.

ity behind the location of the removed plate. If the soil does not fall naturally, it should be removed manually to make room for the new plate to be welded in place. The process of replacing a locally damaged plate of an existing pipe is illustrated in Fig. 10.25.

Although not normally practiced, it is advisable to fill with grout the gap that may remain behind the plate which replaces the locally damaged one.

The disadvantage of repair by internal grouting, besides its excessive cost, is the reduction of the conduit size. As noted earlier, the reduction is of particular concern for culverts.

10.2.4 Shotcreting

Shotcreting with a steel-fiber-reinforced concrete mix has been used successfully to line the insides of the conduits of soil-steel bridges in distress. The lining, which is up to 150 mm (6 in) thick, may cover the complete perimeter of the cross section of the conduit, as shown in Fig. 10.26*a;* alternatively, it may be limited to the damaged zone of the conduit wall as shown in Figs. 10.23(*b*) and 10.26(*b.*)

For the complete ring, shear connectors are not usually provided between the conduit wall and the shotcrete, although their inclusion can certainly increase the usefulness of the additional ring.

The partial shotcrete ring is provided to repair localized damage, such as a section containing bolt-hole tears. For the partial ring, it is important to provide some sort of shear connection between the pipe

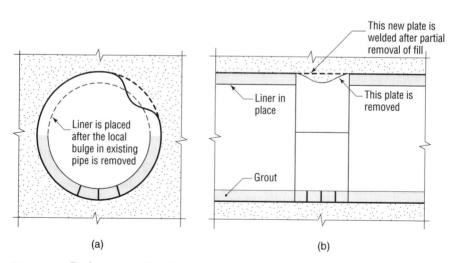

(a) (b)

Figure 10.25 Replacement of locally damaged plates: (*a*) cross section; (*b*) longitudinal section.

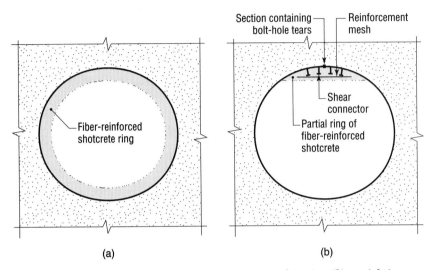

Figure 10.26 Repair by fiber-reinforced shotcrete: (*a*) complete ring; (*b*) partial ring.

and shotcrete. This shear connection may be by shear studs of the type used in composite beams, machine-welded to the pipe after the zinc coating from the galvanized plate has been ground off locally. Shear studs installed on the outside of the pipe can be seen in Fig. 10.28. An alternative to the usual shear stud is a U-shaped bracket which is made out of thin steel plates, and which is attached to the pipe through pins fired by a ram-setting gun. This type of shear connector, which is shown in Fig. 10.27, has been used successfully in several repair works (Ref. 4).

Despite the ability of the fiber-reinforced concrete to sustain fairly large tensile stresses, it is advisable to add a steel reinforcement mesh to the shotcrete ring, especially if it is partial. The location of the reinforcement mesh within the depth of the shotcrete ring depends on the sign of the moments that the repair segment of the conduit wall is subjected to. For example, the segment in the vicinity of the crown is subjected to moments causing tension towards the inside of the conduit. For this case, as shown in Fig. 10.26*b*, the reinforcement layer should be close to the exposed face of the shotcrete. On the other hand, the reinforcement mesh should be close to the conduit wall if the segment under repair is in the haunch area of the conduit. It is recalled that, for this segment, the bending moments induce tension toward the outside of the conduit.

A more detailed account of repair by the partial shotcreting ring is given in Ref. 4.

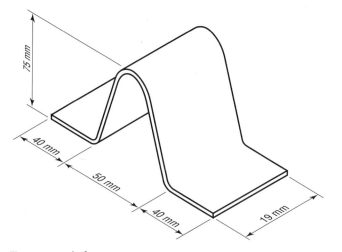

Figure 10.27 A shear connector used in shotcrete repair.

Repair by fiber-reinforced shotcrete can prove economical and effec-tive in many cases mainly because of the fact that it requires no form-work and little preparatory work. Because of its relatively thin layer, the shotcrete ring does not reduce the conduit size appreciably. The repair work by shotcreting can be undertaken even in cold weather, provided that the conduit wall sections to be shotcreted are ade-quately heated. If only the top and side segments are to be repaired, then shotcreting of culverts can be carried out without diverting the stream.

10.2.5 Partial concreting outside conduit

When distress in the conduit wall is limited to only the top segments of the pipe and the depth of cover is shallow, removal of the backfill from above the conduit and adding a layer of concrete to the outside of the pipe may prove to be an economically viable repair method. In this method, the concrete layer is made composite with the pipe through the usual shear studs employed in slab-on-girder-type bridges. These shear studs can be machine-welded readily to the pipe after locally scraping off the zinc layer. Figure 10.28 shows such shear studs welded to the outside of a pipe, the backfill above which was removed for repairs. As shown in this figure, the shear studs are staggered for max-imum efficiency.

Figure 10.29 shows the photograph of a structure, a partial length of whose pipe is reinforced with concrete added to the outside of its top and side segments. This structure, because of its shallow depth of cover, was suffering from peaking deformation leading to frequent bolt-

Figure 10.28 Photograph showing shear studs installed on the outside of a pipe.

Figure 10.29 Photograph showing a partial length of a pipe covered with concrete from outside.

hole tears in a longitudinal seam close to the crown. The addition of a composite reinforced concrete slab not only strengthened the top segments of the conduit wall but also overcame the problem of too shallow a depth of cover.

10.3 Monitoring of Conduit Shape by Photogrammetry

A program for monitoring the shape of a soil-steel structures should accomplish the following:

- Determine the cross-sectional shape of the structure at specified locations and establish whether the structure has deformed from its as-built or design shape.

- Determine, through periodic monitoring, if the deformations have stabilized or are continuing.

- Evaluate other properties that may affect the strength or stability of the structure, such as corrosion, bolt-hole tearing, crimping in the metal plates, and the quality of the backfill material.

This information would ultimately be used in a comprehensive evaluation of the structure.

10.3.1 Traditional monitoring methods

Prior to discussing photogrammetric techniques, other more traditional methods will be reviewed briefly. Traditional monitoring methods have usually consisted of a visual inspection, with actual measurements being taken only if serious signs of distress are observed. Other than the span and rise, geometric measurements are normally limited to selected chord lengths and offsets at specified cross sections, as shown in Fig. 10.30. These dimensions can be related to the curvature of a plate section. The performance of the structure is judged primarily on deformation stability. Excessive flattening of the plate makes it susceptible to snap-through instability. The extent of flattening that a structure can tolerate is not easily defined. Arbitrary limits on the reduction in the midordinate heights have been used to define the severity of the deformations and the remedial measures required (Ref. 5). The changes are usually measured from the design shape, since the as-built dimensions are rarely measured. Being flexible, it is possible that the plates were deformed from the design shape during construction, with little or no subsequent deformations. Only an ongoing monitoring of the structure can confirm that the deformations have stabilized.

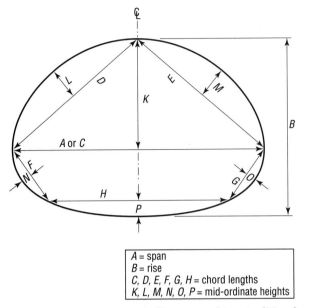

A = span
B = rise
C, D, E, F, G, H = chord lengths
K, L, M, N, O, P = mid-ordinate heights

Figure 10.30 Dimensions typically measured in a traditional monitoring program.

In addition to the limited information that these measurements yield, their accuracy and repeatability are also questionable. In large structures, the tape measure would have to be held in place at the top with a rod, and the midordinate heights read from some distance away. The procedure is further complicated if the structure is full of water, with the measurements being taken from boats. The field measurements are time-consuming and labor-intensive, requiring up to seven persons for one survey.

The measurements are sometimes supplemented with either or both the crown and invert elevations, which provide a more accurate measure of the deformations of the pipe. However, it is not always possible to obtain this information because of difficulties in setting up the survey equipment with suitable sight lines when dealing with large structures full of water.

10.3.2   Monitoring with photogrammetry

In simple terms, in photogrammetry an object is photographed using specialized equipment following set procedures, and measurements are obtained from the photographic images. These measurements and some externally supplied information are used to determine, either analogically or analytically, the location of reference points in the three-dimensional object space. The use of photogrammetry for monitoring soil-steel bridges is an example of close-range, nonterrestrial photogrammetry, as opposed to aerial photogrammetry used for mapping purposes.

Some of the advantages of photogrammetric monitoring are as follows:

- Taking photographs is less labor-intensive than taking direct measurements in the field.

- It is no less accurate than direct measurements, and probably more accurate in large or difficult-to-access structures.

- A large amount of geometric data can be extracted from the photographs, and the cross sections can be defined in as much or as little detail as required for the subsequent analysis.

- Because the geometry of the structure is frozen in time, additional measurements can be taken from the same photographs at a later date, if required, or questionable values checked, without having to repeat the field work.

- In addition to logging changes in geometry, the photographs from subsequent surveys provide objective documentation of other characteristics which may affect the capacity of the structure, such as the extent of cracking, corrosion, etc.

- State-of-the-art systems do not require special skills or training, either for the photography or the analysis of the photographs.

10.3.3 Fundamentals of photogrammetry

Photogrammetry can be divided into three fundamental components:

1. the actual photography
2. taking measurements from the photographic images
3. analysis to convert the resultant two-dimensional image coordinates into three-dimensional object coordinates

In general, the three components are part of an integrated system in which the equipment and procedures of the various stages are interdependent. Although the science of photogrammetry has evolved significantly over the years, with corresponding changes in the equipment and procedures, the fundamental concepts remain the same. It is not within the scope of this section to discuss those concepts in any detail, nor to document the evolution of photogrammetry to its current state. The reader is referred elsewhere for that information (Ref. 1). This section will only deal with the fundamentals of photogrammetry in general terms, and will discuss specific equipment and procedure as they affect this particular application.

Photogrammetry is based on the same principles as stereoscopic vision, whereby an object viewed through two eyes appears in three dimensions. If an object is photographed from two positions, each point on the object can be described mathematically as the intersection point of two lines, each passing through the point itself and the projection centers of the camera(s) at the time of the exposures, as illustrated in Fig. 10.31. The determination of these intersection points requires that certain constants be known beforehand. The orientations can be grouped into three categories:

1. *Internal orientation* defines the internal characteristics of the camera: the focal length, the location of the principal point on the image plane (i.e., the point through which the optical axis passes), and the orientation of the image plane to the optical axis, which is preferably perpendicular.

2. *Relative orientation* defines the relative positions of the cameras when the two photographs were taken, namely, the distance apart and the relative directions of the optical axes.

3. *Absolute orientation* defines the scale of the object and its location in an external reference system.

In addition, correction factors may be needed in order to compensate for distortions which can arise from a variety of sources, such as imper-

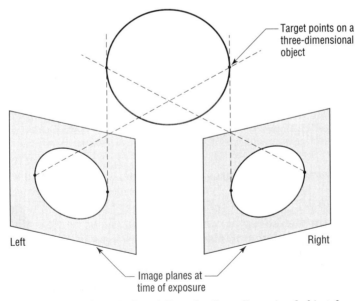

Figure 10.31 Mathematical modeling of a three-dimensional object from two-dimensional images.

fections in the lens, nonplanarity of the film in the camera, instability of the film material due to temperature or humidity, etc. Similar distortions can also be introduced whenever a print or enlargement is made from the negatives.

Traditional photogrammetry. In traditional photogrammetry systems, sophisticated equipment and strict procedures are used to establish most of the orientation constants and to eliminate, or correct for, distortions during the photography stage. The constraints associated with these systems often make them unsuitable for an application such as monitoring soil-steel bridges, either because the rigid photographic procedures cannot be adhered to under the field conditions, or the systems are not cost-effective because of the cost of the equipment and the need for highly skilled operators.

Recent developments. Advances in the science of photogrammetry over the years have resulted in improved analytical techniques which, instead of requiring all the orientation constants to be established during the photography stage, solve for them during the analysis. The procedures and equipment for taking the photographs and taking measurements from the images are thereby greatly simplified. Methods have also been developed which compensate for distortions and other errors, thereby further reducing the need for sophisticated and expensive equipment. The end result is that photogrammetric systems are available which are more economical, suited to wider applications, and do not require highly skilled operators. The use of simplified equipment and procedures may, however, result in reduced accuracy as compared to traditional photogrammetry systems. Considering the flexibility of the metallic shells of soil-steel bridges and the large deformations they can withstand, the accuracy of the system used is nevertheless found to be adequate for all practical purposes.

The system which has been used in a particular monitoring program is the Elcovision 10 system, developed by Wild Heerbrugg of Switzerland (Ref. 5). The system uses a calibrated and modified Leica 35mm camera, which is a high-quality camera but relatively inexpensive when compared to the large-format, true metric camera used in traditional photogrammetry. A variety of interchangeable lenses are available. Calibrations set the internal orientation constants for each camera/lens combination and provide correction factors to compensate for lens distortions.

The camera is modified by the inclusion of a reseau plate in the camera body which superimposes a 5×7 grid of crosses onto the photographic image, as shown in Fig. 10.32. The spacing of these crosses on the reseau plate is calibrated precisely. The spacing of the crosses on the photographic image defines any distortions which have occurred after the light has passed through the lenses, due to nonplanarity of the film in the camera, instabilities in the film or print paper, etc. It also compensates for distortions introduced during the printing and enlarging processes. This allows measurements to be taken from enlargements, say 200×250 mm (8×10 in), rather than directly from the film negatives as is done in most traditional photogrammetry systems. This in turn eliminates the need for high-precision measuring devices such as mono- or stereo-comparators which are capable of taking measurements from the negatives to an accuracy of a few microns. Instead, measurements are taken from a pair of enlargements using a digitizing tablet with pointing accuracies in the neighborhood of ± 0.13 mm (± 0.005 in). The digitizer is attached to an AT-compatible personal computer on which the proprietary software runs. The program calculates three-dimensional object coordinates from the two-dimensional image coordinates. These can be fed into a CAD program for detailed plotting or manipulation, or used as input to a structural analysis program. In all, the entire system is relatively inexpensive and simple to operate, requiring no advanced skills or training. The actual steps involved in the photography and subsequent analysis of the photographs are subsequently discussed in detail.

10.3.4 Elements of the monitoring program

Although the fundamentals remain unaltered, procedures have been established which are suited to the monitoring of soil-steel bridges,

Figure 10.32 Photograph showing grid marks.

ensuring efficiency and accuracy. These include steps to prepare the structure, as well as the photography and analysis. Preparation of the structure is straightforward; it includes selecting cross sections to be monitored and installing targets to define those sections.

Target selection.   The targets are the points which will be digitized in the photographs. Although any identifiable point in the photographs can be used as a target point for digitizing, it should be small and highly visible for optimum results. For example, nuts or bolt heads could be used as target points, but if left unmarked they can be difficult to distinguish from the metal plate and from other nuts. Even if they are marked by spraying them with fluorescent paint, they are large enough to make it difficult to digitize precisely the same point on the nut in a pair of photographs. Small errors in digitizing object points can have a very significant impact on the overall accuracy of the final coordinates. It is therefore recommended that small, high-contrast targets be attached either to the nut or directly onto the corrugated metal plate. A further advantage of having unique targets is that the same points are digitized in subsequent surveys, and the coordinates are compared so as to identify and locate a change in geometry quickly.

Flash photography is required because of the poor lighting conditions inside the structures. It was found after several trials that the best targets were made from the reflective pressure-sensitive material used for lettering highway signs. The higher the reflectivity, the better. The material is available in sheets from which 6-mm- (0.25 in) diameter circles are punched, which are small enough for precise pointing on the digitizer. These are applied directly onto the nut or metal plate after the surface has been thoroughly cleaned and degreased. It is recommended that a small circle of flat paint be applied to the surface and allowed to dry completely before the target is applied. The paint serves two purposes. First, it provides better contrast for the reflective target than does the galvanized metal, resulting in more precise digitizing, as can be seen in Fig. 10.33. Second, it facilitates the location of specific targets in different photographs, thus reducing errors when digitizing. Different colors can be used at different points along the section to further simplify the location of specific targets. Flat black and fluorescent red or orange were found to work well, being clearly visible and providing good contrast for the targets. After the targets have been applied, a finishing coat of nonglossy, clear acrylic is also recommended as a means of protecting them.

Target locations.   The cross sections to be targeted can be located at regular spacings throughout the length of the conduit or just in the

Targets

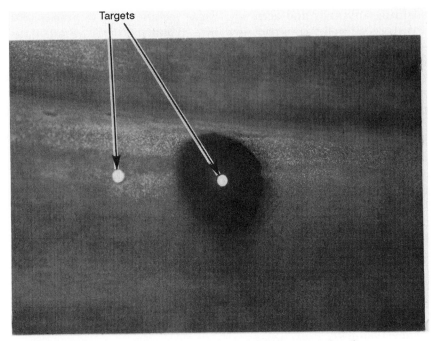

Figure 10.33 Photograph showing targets on painted and unpainted surfaces.

areas of concern. Section spacing from 3 to 6 m (10 to 20 ft) has been used, depending on the size of the structure. Most soil-steel bridges have continuous circumferential seams at fairly regular spacing, and the exposed nuts would seem to be ideal locations for the targets. However, there are several arguments for not using circumferential seams for target sections:

- Some structures have rib stiffeners at the circumferential seams, and even in those that do not, the overlap of the plates has some stiffening effect. Because of such local stiffening, the deformations under the seams are likely to be smaller than at sections away from the circumferential seams.

- Although the nut provides a flat surface on which the target can be applied, the orientation of the nut may be such that the target is visible from one side of the structures but not the other. For the photogrammetric analysis to be valid, it is important that all the targets be clearly visible in two photographs taken from different positions.

- Those nuts which coincide with both a longitudinal and circumferential seam may be obscured by other bolts in the longitudinal seam when viewed from certain angles.

The preferred option is to attach the targets directly onto the corrugated metal plate at sections midway between circumferential seams.

The photographs are taken looking along the length of the structure as shown in Fig. 10.34, and the targets should be attached so that they are facing the camera. The optimal location is near the inside crest, approximately one-third the distance between the crest and the neutral axis, as shown in Fig. 10.35. This positioning ensures that the target is not obscured by other corrugations when photographed from the near side, and the orientation provides sufficient reflection from the flash for most camera locations.

The targets should be well dispersed around the portion of the cross section above the waterline or not covered with debris. A minimum of three, preferably four to five targets should be placed on each plate segment, i.e., between each longitudinal seam. This is to facilitate the reconstruction of the cross section from the generated coordinates by passing arcs through the target points from each plate segment, which requires a minimum of three points. Although individual plates may flatten or become more curved under load, changes in curvature and discontinuities in slope are most likely to occur at or near longitudinal

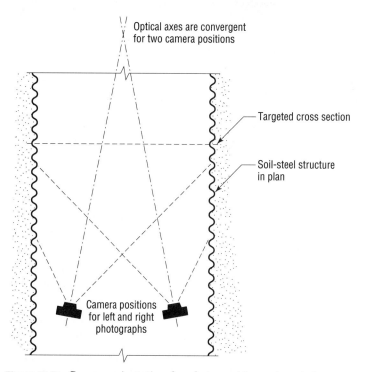

Figure 10.34 Camera orientation for photographing a targeted cross section.

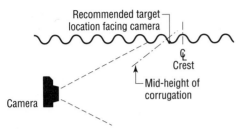

Figure 10.35 Installing targets on the corrugated steel plate.

seams, either by design or as a result of poor fitting joints, crimping, or cracking. If there is a noticeable kink within a plate section, additional targets should be placed on either side of the kink to better model the shape of that plate. An additional target can be placed near each longitudinal seam in the section, preferably with a different-colored background paint, so as to distinguish each plate segment in the photographs, as shown in Fig. 10.36. Once the targets have been installed, each section should be labeled with paint or permanent marker for proper identification.

Photography.   Prior to each photo session, magnetic signs should be attached at each section to identify them in the photographs, taking care not to obscure the targets. If necessary, the targets should be brushed clean. Even a thin layer of frost will reduce their visibility in the photographs. Another sign may be desirable to identify the actual

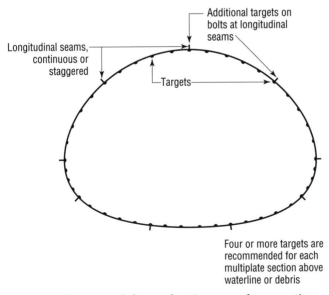

Figure 10.36 Recommended target locations around cross section.

structure. The use of a databack on the camera is recommended so that the date of the survey can be recorded directly on the photograph.

Scaling devices.   In order to establish the size of the section from the photographs, objects of known dimensions must be visible in them. Standard surveying rods were found to be unsatisfactory because the continuous markings make it difficult to distinguish precise points for digitizing. As an alternative, aluminum angles 1 m in length with a single reflective target at the center of one side were fabricated. Individual segments are bolted together to form scaling rods of desired lengths, as shown in Fig. 10.37. It is recommended that the longest lengths practical be used, since accuracy improves with the length of the scaling device. A minimum of two, preferably three, rods should be used. These are positioned in the structure in the vicinity of the cross section being photographed, oriented in different directions with reflective targets facing the camera. In dry conduits, they can lie on the ground and be propped up against the sides. In conduits with water, they may have to be suspended from hooks installed during the initial targeting. In any case, they must be secured so they do not move between the two photographs of each section.

Camera position.   As mentioned earlier, the analysis requires two photographs of the cross section from different positions: one from the left and one from the right side. The further apart the positions, the better,

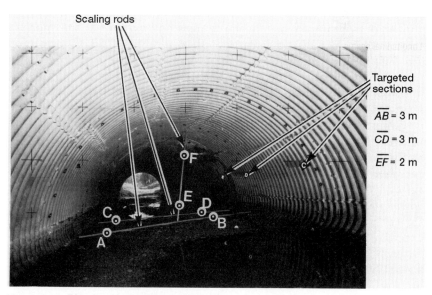

Figure 10.37   Photograph showing scaling rods.

but if the camera is too close to the side the reflection of the flash may obscure the targets. In a small structure, it may be necessary to crouch low so that the targets on the top are visible. The camera should be oriented as shown in Fig. 10.34, so that the optical axes for the two images are convergent and not divergent. The camera can be handheld, but should be kept approximately horizontal with a similar tilt from the vertical in both pictures.

The entire cross section should be visible in both photographs using a 24-mm wide-angle lens. It is best to take the photographs from far enough back so that the section is not near the outer edges of the frame, since this is where the maximum lens distortions occur. At these distances, the section should be in focus with the focusing ring set at infinity. It is recommended that all photographs be taken with the focusing ring locked on infinity since different calibration factors apply for different settings, and there is the risk that the focus setting may not be recorded properly.

Although more than one section may be clearly visible in a pair of photographs, accuracy decreases with distance from the camera. For the time and expense involved, it is recommended that each section be photographed separately, moving the scaling devices and camera into position for each.

Exposure.   As previously stated, flash photography is required. A powerful flashgun should be used, with a wide-angle dispersion attachment. For small structures, having spans less than 3 m (10 ft), one flashgun on the camera is sufficient. For larger structures, two flashguns may be required, one on the camera and another at the opposite side attached by a PC synch cord, as shown in Fig. 10.38. Determining the flash setting and aperture for proper exposure is often done by trial and error, and depends on the size of the structure, the light intensity from outside, the surface condition of metal plate, the presence of water, etc. In general, it is better to overexpose than underexpose, since a sufficient number of reseau crosses must be visible in both the photographs to establish the distortion correction factors. In an underexposed photograph, there may not be sufficient contrast between the reseau marks and the background to digitize them accurately, as can be seen in Fig. 10.39. For initial photographic sessions, it was found that setting the flashgun at full (manual) and the aperture between $f4$ and $f8$ gave reasonable results. However, a variety of exposure conditions, i.e., flash and aperture settings, should be tried for each structure until an optimum range is found. The trial settings should be noted and the frame number recorded on the databack for later reference.

Figure 10.38 Photograph showing the camera and two flashguns.

Figure 10.39 Poor resolution of grid marks on an underexposed photograph.

Film and processing.   Color film is recommended over black and white for ease of locating targets on the prints, particularly if different-colored paint marks are used to divide the cross section. It is not necessary to use professional grade or high-speed film. A good quality 100 or 200 ASA film will result in adequate resolution for target identification.

To ensure satisfactory results, the exposed film should be processed immediately, and the prints inspected. The best left and right photograph for each section can be identified for enlarging. Although the reseau mark transformation does correct for distortions introduced during processing and enlarging, better accuracies are obtained if these distortions are minimized by the use of quality processing and materials.

10.3.5   Photogrammetric analysis

Internal orientation.   The analysis begins by identifying the camera, lens, and focal length combination(s) used for a pair of photographs. The software has stored the internal orientation parameters and calibration for each combination. The left and right enlargements are mounted on the digitizing table side by side. All digitizing is done for one point or target at a time, first in the left photograph, then in the right. The reseau marks are first digitized, which the program uses to calculate distortion correction factors. Although a minimum of nine marks is sufficient, it is preferable to use all the reseau marks which can be distinguished in both photographs.

Relative orientation.   The next phase is the reconstruction of the relative orientation. A number of points are digitized in both the photographs. These can be targets or other distinct points on the structure, but not the reseau marks on the photographs. A minimum of five points are required, although thirty or more are recommended. By using more than the minimum required, the program calculates (in addition to the relative orientation parameters) measures of accuracy, and identifies potential error points (i.e., points which were incorrectly digitised or mismatched in the two photographs). The points used should be well dispersed over the photographs and should include points in the cross section being analyzed and in other sections as well. Figure 10.40 shows a selection of target points suitable for relative orientation. Only the left photograph is shown; the same targets must be visible in the right photograph as well.

Absolute orientation.   In this stage the size, location, and orientation of the structure are determined in a user-defined reference system. This

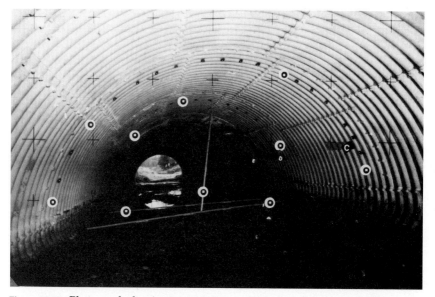

Figure 10.40 Photograph showing target points suitable for relative orientation.

can be done by digitizing three or more control points common to both photographs. These points should be permanent benchmarks with known coordinates in an external reference system. The advantage of this approach is that the coordinates of all the targets at all sections would be calculated in the same fixed-reference system, allowing deflection profiles as well as cross-sectional deformations to be determined. Unfortunately, the control points would have to lie outside the structure so as not to be influenced by its deformations. This means they would be confined to the small portion of the photographs showing the end of the tunnel, and they would be a considerable distance from the camera. As will be discussed later, these conditions generally yield poor accuracies and are not recommended.

The preferred approach is to define a local reference system for each pair of photographs. Two targets near the spring line are used to define the direction of the positive x axis, and a third target near the crown defines the xy plane as shown in Fig. 10.41. The direction of the positive z axis is out of the plane of the page by the right-hand rule. The section is located in this local reference system by assigning known or assumed coordinates to a target. For example, a target near the crown could be given coordinates $x = 0$, $y = 100$ or the actual elevation if it is available from a survey, and $z =$ the distance from the downstream end of the structure. If the same targets are used to define the reference system in subsequent surveys, then any change in cross-sectional

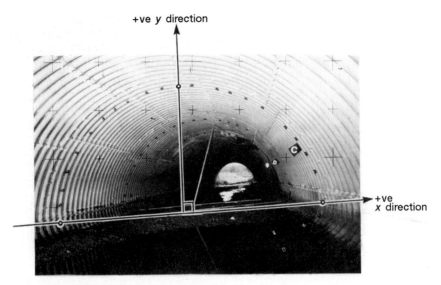

Figure 10.41 Local reference system.

shape can be readily detected in that reference system. It is not possible to calculate absolute displacements of target points or of one cross section relative to another, unless the local reference systems are tied into a global reference system.

The last step in the absolute orientation is to establish the scale of the structure in the local reference system. Scaling is performed by digitizing points on the scaling rods and entering the corresponding distance between them (Fig. 10.37). A single known distance is sufficient, but additional known distances provide some degree of redundancy and can lead to the identification of possible errors. Scaling distances in different directions further improves the accuracy.

Digitizing and postanalysis. Having completed the orientations, three-dimensional coordinates are then calculated for all target points digitized in both photographs. These can be analyzed in a number of ways:

- The coordinates for individual targets can be compared from one survey to the next to detect deformation.

- Distances between target points can be calculated and compared to the distances from other surveys or to distances measured directly.

- Two-dimensional plots of the cross section can be generated from the projection of the coordinates onto the xy plane. The Elcovision program can generate simple straight-line plots, or the coordinates can be transferred into a CAD program for more complex plotting. Figure 10.42 shows the cross section of what was originally a circular pipe. Each plate segment is represented by a separate arc, showing the radius of curvature of each arc and the angular discontinuity between plate segments as calculated by the CAD program. These provide a quantitative and visual measure of the distortions which have occurred, either from the design shape or since the last photogrammetric survey of that structure. Plots can also be superimposed to show graphically the change in cross-sectional shape between surveys or along the length of the conduit.

■ The coordinates can be used to define the cross section in a structural analysis program (e.g., finite element analysis) which would calculate the actual force effects and allow the level of safety of the structure to be determined.

10.3.6 Accuracy

During the evaluation of the Elcovision 10 system for monitoring soil-steel bridges, accuracy was found to increase with operator experience and refinements in methods and equipment. The foregoing procedures incorporate those refinements.

Accuracy was measured by the residuals between distances calculated from the Elcovision coordinates and distances measured directly on the structure. A relatively small structure having a span of approx-

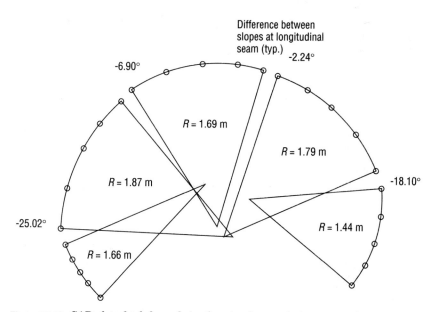

Figure 10.42 CAD plot of a deformed circular pipe from a photogrammetric survey.

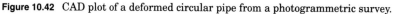

imately 4 m (12 ft) was used to ensure confidence in the measured distances. Several targeted cross sections were digitized from a single pair of photographs. Mean residuals in the xy plane of the cross section containing the scaling devices ranged from 2 to 7 mm (0.080 to 0.276 in), while the mean residuals in the cross sections without the scaling devices ranged from 30 to 80 mm (1.18 to 3.15 in). Mean residuals in the longitudinal z direction, i.e., between sections, ranged from 20 to 40 mm (0.787 to 1.575 in).

One reason for these variations is the short base length between the two camera positions necessitated by the geometry of these structures. As shown in Fig. 10.43, this leads to an acute angle between the sight lines connecting a target point and its image in the two photographs. A small error in digitizing the point in one of the photographs will result

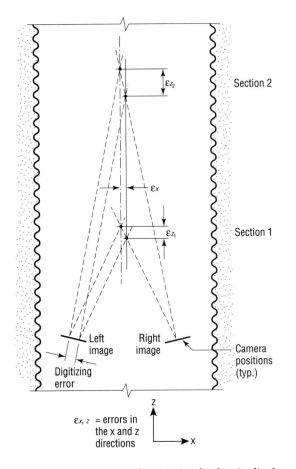

Figure 10.43 Error magnification in the longitudinal

in a small error in the xy coordinates (ε_x) but could translate into a large error in the z coordinate, more so as the distance from the camera increases (ε_{z1}, ε_{z2}). For this reason, it is recommended that only one cross section be analyzed from each pair of photographs, that the scaling devices be positioned near that cross section, and that the targets used to establish the relative orientation and to define the local reference system also be selected from or near the targeted section.

References

1. American Society of Photogrammetry, *Manual of Photogrammetry,* 4th edition, Falls Church, Va., 1980.
2. Bakht, B., and Agarwal, A. C., "On Distress in Pipe-Arches," *Canadian Journal of Civil Engineering,* 15(4), pp. 589–595, 1988.
3. Cowherd, D. C., and Degler, G. H., *Evaluation of Long-Span Corrugated Metal Structures,* Bowser-Morner Associates, Inc., Dayton, Ohio, 1986.
4. Ramotar, J. C., "Culvert Seam Failure and Repairs," unpublished report of Alberta Transportation, Edmonton, Canada, 1986.
5. Wild Leitz, *Elcovision 10 Handbook,* Wild Heerbrug Ltd., Heerbrug, Switzerland, 1988.

Principles of Soils Investigation

Cameran Mirza

Strata Engineering Corporation
Don Mills, Ontario, Canada

A.1 Introduction

Soils, particularly those below and around the conduit, play a significant role in the performance of the soil-steel composite structure. Therefore, soil investigation is essential to ensure the economy and safety of the structure.

The investigation of soils for soil-steel bridges may be divided into two logical components:

1. Investigation of the foundation soils below the conduit
2. Investigation of the soils to be used as bedding and backfill to the conduit

This appendix describes the principles for such investigations and leaves the details to manuals of standard practice.

A.2 General Principles

All soil-steel bridges larger than about 3 m (10 ft) in span should be subjected to a geotechnical site investigation. The chief purpose of such an investigation should be to determine the subsurface conditions prevailing below the proposed conduit and to assess whether or not such conditions are critical to the safety of the structure.

The key elements of such an investigation should include:

- if the stability of the structure and associated fills or cuts is affected adversely by the proposed construction

- if the immediate, short- and long-term settlements of the structure and its components can be tolerated safely
- the time rates of such settlements

Stability is a function of the resisting characteristics of the soil when a load is imposed upon it. Settlement is a characteristic of the previous geological history of the soil. Time rate of settlement is a function of the permeability of the soil.

A.3　Methods of Investigation

There are several methods of investigating foundation conditions at the site of a proposed structure. They range from an evaluation of previously available, reliable information to actual drilling and sampling of boreholes to identify the stratigraphic sequencing of various soils below the invert of the structure.

Generally speaking, the site investigation should start with an evaluation of the prevailing geology of the area, with special reference to local conditions at the proposed crossing site itself. National and Provincial geological maps for a variety of uses and cases are available at small cost, and should be routinely consulted, regardless of the extent or intensity of the proposed investigation. With an understanding of the geological setting, a site investigation program can then be tailored to meet the anticipated geological conditions and the requirements of the structure.

As a rule, stability considerations include an assessment of the bearing capacity of the structure and its associated fill, as well as stability of the natural soils either below or beside the conduit. Therefore, the investigation must be designed to satisfy the information requirements for such assessments. The ultimate purpose of any site investigation should be to characterize the soils below the proposed conduit to depths sufficient to satisfy bearing capacity, stability, and settlement information requirements.

As opposed to footings for conventional structures, for which the stresses dissipate rather rapidly with depth below the footing, the stress regime below fills associated with soil-steel structures is widespread, in that very little stress reduction occurs with depth. Therefore, lack of sufficient depth of investigation could lead to the overstressing of a weak soil layer at depth, causing instability of the fill and/or undesirable settlements of the conduit, with consequent influence on the structural behavior of the conduit itself.

As a rule of thumb, the minimum depth of a site investigation should be twice the proposed height of fill or depth of cut; the depth should be still greater if unsuitable soil conditions are found to be present at or within that depth.

A.4   Soil Types

For practical purposes, foundation soils may be divided into three broad categories:

- Inorganic cohesive
- Inorganic noncohesive
- Organic

Inorganic cohesive soils can be further subdivided into normally consolidated and overconsolidated.

Sites with organic or normally consolidated soils below the conduit require special attention and care. Such soils offer little resistance to imposed loadings and suffer from extremes of settlement.

Inorganic noncohesive soils include sands, silts, and gravels. Generally speaking, such soils are safe for most soil-steel bridge installations from the bearing-capacity and fill-stability point of view, but they do require special attention with respect to erosion, piping, and frost heaving, especially in hydraulic applications such as stream crossings.

Overconsolidated clayey soils must be investigated for a characteristic known as *residual strength*. Initially, such soils show good capability to support the proposed fill or cut safely. However, in time, their resistance characteristics change to a condition known as the residual value of the shearing strength. Many cut-slope failures in stiff clays have been attributed to this change in strength characteristics after a number of years. Residual strength is generally not a problem where fills are concerned.

The settlement of silts, sands, and gravels is generally acceptable for most soil-steel bridge applications and need not be calculated to the extent necessary for rigid structures. However, the settlement of organic and normally consolidated clayey soils can be a major concern for soil-steel bridges and must be calculated to allow properly for the anticipated settlements in the design of the invert profile of the structure.

The time rate of settlement is essentially a function of the permeability of the soil type. Sands and gravels, having excellent permeability characteristics, generally settle immediately. Silts, being intermediate in permeability between sands and clays, also generally settle within a short time.

However, clays and organic soils settle not only immediately upon application of load by elastic compression, but also continue to settle over a longer period of time. Clays undergo a settlement known as *primary consolidation* and, unless seriously overstressed, will not undergo secondary compression after or during completion of the primary, when the excess water in the pore spaces of the soil fabric has drained away.

By contrast, organic soils invariably suffer from almost immediate secondary compression, due to the inherent structural weakness of the soil components. This can contribute not only to large settlements, comparable to those of softer clays, but also to ongoing settlement by a creep-type behavior.

A.5   Site Investigations

The purpose of the site investigation for soil-steel bridges is to elicit such information as will enable the foundation soils to be properly classified, and for an engineering assessment of their behavior, under the proposed structural requirements, to be suitably made.

Details of borehole drilling or test-pit evaluations, borehole sampling, in situ testing, and laboratory testing of recovered soil samples are provided in many manuals and guides, as well as in elementary textbooks on soil mechanics. For Canadian practice, a good reference manual is the *Canadian Foundation Engineering Manual*. This manual is a Canadian Geotechnical Society publication.

To assist the geotechnical engineer with the site investigation, the structural designer or the owner should provide the following minimum information:

- site location and purpose of structure
- span, length, and shape of structure
- height of fill above crown, or maximum depth of cut
- site access and property clearance status

It is good engineering practice for the structural engineer and/or owner never to indicate borehole or test-pit locations.

The competent geotechnical engineer would know from the preceding specifications and the geology of the site what type of investigation to undertake, and to what level of intensity in terms of coverage and depth.

Unfortunately, the practice of inviting cost quotations for geotechnical site investigations normally results in a conservative assessment of site geological and soil conditions for soil-structure applications, with the end result that the few hundreds of dollars saved by this "bidding" procedure are more than offset by substituting a more costly conventional bridge, usually supported on piled foundations, at an extra cost of tens of thousands of dollars.

Careful site investigations, particularly in softer clays, require careful drilling and sampling, in situ testing, and further careful laboratory tests to determine the shear strength and consolidation characteristics of the clay. All of these are time-consuming and costly.

To ensure that the proposed crossing or grade separation structure is the least-cost alternative for the soil conditions prevailing, it would be prudent for cost-subsidizing agencies to discourage such bidding practices, particularly by municipal owners.

A.6   Materials of Construction

Materials of construction refer to those soils which are to be used immediately below and around the proposed conduit.

Careful selection and application of material is a key to achieving safety and economy of design. For soil-steel bridge applications, only the noncohesive soil types should be used.

Silts and fine sands, although noncohesive, are considered to be "frost-susceptible," and can exert extremely high pressures when frozen and heaving. Failures of soil-steel bridges have occurred when such soils, present below the invert, have frozen in winter, and in springtime have continued to heave through attraction of water to the freezing front. The basal plates and inverts of conduits have buckled locally under such heaving forces, allowing the free flow of water below the invert, with consequent uplifting of the entire structural conduit in a hydraulic mode of failure. For this reason, the materials usually considered acceptable for bedding and backfilling are sands and gravels.

Some types of clayey soils (such as some Canadian glacial tills) may be used outside the engineered fill zone of the soil-steel structure, provided that they are relatively incompressible after placement.

Noncohesive soils are amenable to compaction, as are other types, including some clayey soils. Generally, such soils can be compacted to a maximum value for any given moisture content, at a given level of compactive energy per unit of soil volume. The standard of compaction is generally the Proctor standard, which is discussed in Chap. 7. The compaction achieved in the field is measured by comparing the field density of the soil with that obtained in the laboratory, at the same moisture content, under a given energy of laboratory compaction.

The geological study of the area will usually have revealed the potential sources for granular noncohesive materials suitable for bedding and backfilling; alternatively, the local governmental agencies and municipalities may be contacted to obtain a listing of the potential or commercial sources for such materials. To achieve economy in design, every effort must be made to utilize the locally available soil. Whenever modification of the local soil is proposed by mixing with chemical or similar additives, a thorough research program must be conducted to ensure that the soil is not being rendered unsafe in other modes.

The strength and resistance characteristics of compacted, noncohesive soils increase to a maximum value beyond which "dilation" can occur, leading to a loosening of the soil fabric. Hence, care is required when compacting soils, to ensure that they are not overly compacted to the point of dilation. Standard methods of compaction and material quality control are available in manuals and applicable Standards.

As a rule, the engineered fill zone around the conduit must not contain soil that is "plastic" (i.e., cohesive) nor a soil that contains more than about 10 to 12 percent by dry weight of "fines" that are smaller than 75 micrometers in grain size. It is believed that noncohesive soils containing fines greater than about 10 to 12 percent will be frost-susceptible.

The materials required to ensure a positive cutoff against hydraulic flow are, of course, the opposites of those required for bedding and backfilling. The soils must be cohesive enough to exhibit low permeability, and yet should be strong enough to withstand the imposed stresses during construction and service.

Flexural Behavior
of Bolted Connections

Leonid Mikhailovsky

Ministry of Transportation of Ontario
Downsview, Ontario, Canada

D. J. Laurie Kennedy

University of Alberta
Edmonton, Alberta, Canada

B.1 Introduction

As discussed in Chaps. 3 and 4, the conduit wall of a completed soil-steel bridge is subjected to predominantly compressive forces. The attendant bending moments are relatively small because of the small flexural rigidity and capacity of the corrugated plates commonly used, i.e., the ones with a 152×51 mm (6×2 in) corrugation profile. Details of this profile are given in Chap. 1, Fig. 1.5.

When the bending moments applied to a conduit wall exceed its flexural capacity, a hinge is formed. It is generally considered that the formation of such plastic hinges at strategic locations around the circumference of the pipe is responsible for keeping the bending moment regime of the pipe at a low level. Because the bending moments in the conduct wall are generally not accounted for explicitly in the design calculations, the joints at longitudinal seams are also designed for direct thrust only. The axial capacities of conventional joints with different bolting arrangements can be obtained from Chap. 5, Fig. 5.9, and those for joints with slotted holes from Chap. 8, Fig. 8.7

Recently, experimental and analytical investigations were undertaken to study the flexural behavior of bolted joints of 152×51 mm (6×2 in) corrugated plates (Refs. 1 and 2). These investigations have shown that the bolting arrangement does not have a significant influence on the flexural capacity of a joint; however, its influence on the ductility and bending deformations of the joint is highly significant. These investigations show that only some of the bolting arrangements commonly employed in soil-steel bridges should be used.

Some details of the aforementioned investigations are given in this appendix, together with the recommendations for preferred bolting arrangements.

B.2   Geometry of Bolted Joints

As discussed in Chap. 1 and illustrated in several figures (e.g., Fig. 1.16), the pipe of a soil-steel bridge is formed by lap-jointing curved, corrugated plates at longitudinal and circumferential seams. At circumferential seams, there is not a significant transfer of loads across the joints; accordingly, the plates are joined at these seams by bolts in only one row, as illustrated in Chap. 1, Fig. 1.17. By contrast, at longitudinal seams, the joint transmits not only thrusts but also bending moments. To enable it to transmit bending moments, the joint must have at least two rows of bolts, which is the current practice. Joints at longitudinal seams are dealt with in this appendix without further reference to the orientation of the seams.

Bolted joints can have as few bolts as shown in Fig. B.1a, b; in this bolting arrangement, which is referred to as arrangement 1 in this book, the bolts are placed on all the ridges in one row and in all the valleys in the other. The maximum number of bolts occur in bolting arrangement number 3, which is illustrated in Fig. B.1c; in this case, every ridge and valley of each row contains a bolt. In bolting arrangement number 2, which is shown in Fig. B.1d, e, and f, every ridge and valley of one row contains a bolt, and the other row contains bolts either in all valleys or on all ridges.

As was explained in Chap. 4, a desirable pattern of bolts for arrangement number 1 is that in which the bolts closer to the visible edge of the mating plates are placed in valleys rather than on ridges. This desirable bolting arrangement is shown in Fig. B.1a. It is worth noting that the concept of desirable and undesirable bolting arrangements was proposed jointly by the first-named author of this appendix and T. Belke, when the former was with Alberta Transportation and Utilities.

B.3   Failure Loads of Bolted Joints

The datum for the behavior of the joints of corrugated plates of a given thickness is the moment-curvature plot for the corresponding plate without the joint. Such a plot is obtained theoretically and verified experimentally for a 5-mm- (0.197 in) thick corrugated plate (Refs. 1

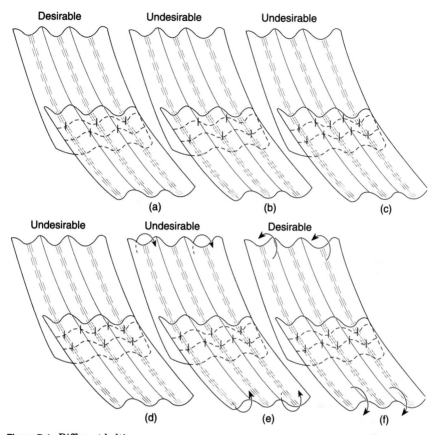

Figure B.1 Different bolting arrangements.

and 2); it is reproduced in Fig. B.2. It can be seen in this figure that the flexural behavior of the corrugated plate is highly ductile and very close to the idealized elastic-plastic behavior. For it to be compatible with the flexural behavior of the parent plates, a bolted joint must have not only similar failure moments, but also a similar mode of failure.

A number of simply supported corrugated plates incorporating, respectively, all the bolting arrangements currently employed in practice were tested in the study described in Refs. 1 and 2; these tests were carried out under a two-point loading scheme in which the joint is in a state of constant moment. Details of the tested joints are given in Table B.1, along with the failure loads and the ratios of these to corresponding failure loads, M_p, of plates without the joints. It can be seen in this table that the failure loads of all the joints are fairly close to the failure loads of the corresponding parent plates. In only a few cases is

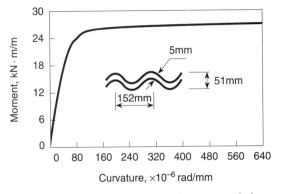

Figure B.2 Moment-curvature plot of a corrugated plate.

the failure load of the joint higher, albeit slightly, than the failure load of the parent plate.

In light of the earlier discussion with respect to the beneficial effect of plastic hinges around the pipe, it is a desirable attribute of the bolted connections that their failure loads are likely to be somewhat lower than those of the parent plate; the desirability of this attribute should nevertheless be qualified by the requirement that the load-deformation characteristics of the joint should be ductile and without the consequence of distress that can grow with time.

The effects of the various parameters on the moment-rotation characteristics of the bolted joints are discussed subsequently.

TABLE B.1   Ultimate Moment Capacities of Bolted Joints

Test no.	Plate thickness, mm	No. of bolts per corrugation pitch	Bolting arrangement of Fig. no.:	Desirable/ undesirable	Bolt torque N·m	Ultimate moment/meter width, M_u	$\dfrac{*M_u}{M_p}$
1	3	2	B.1a	desirable	250	15.6	0.961
2	5	2	B.1a	desirable	500	25.5	0.959
3	5	2	B.1a	desirable	250	23.9	0.898
4	5	2	B.1a	desirable	50	22.7	0.853
5	5	2	B.1b	undesirable	250	23.6	0.889
6	5	2	B.1b	undesirable	50	22.5	0.847
7	7	2	B.1a	desirable	250	38.7	1.079
8	7	3	B.1d	undesirable	250	37.0	1.030
9	7	3	B.1e	undesirable	250	36.3	1.011
10	7	3	B.1e**	undesirable	250	35.3	0.981
11	7	3	no figure	undesirable	250	33.5	0.933
12	7	3	B.1d	undesirable	250	34.6	0.964
13	7	4	B.1c	undesirable	250	35.4	0.987

* M_p = analytical plastic moment of the corrugated plate.

** Bolts with washers.

B.3.1 Effect of bolting arrangement

The effect of desirable and undesirable bolting arrangements on the failure mode of the joint is illustrated in Fig. B.3a with respect to 5.0-mm- (0.197 in) thick plates joined according to the schemes of Fig. B.1a and b; in these schemes there are two bolts per pitch of the corrugation. It can be seen in Fig. B.3a that the joint with desirable bolting arrangement has the same ductile behavior as that observed in the parent plate and shown in Fig. B.2. The joint with the undesirable bolting arrangement has practically the same failure moment as the other joint, but exhibits a lack of ductility. This lack of ductility is caused by the formation of bolt-hole tears of the kind shown in Chap. 10, Fig. 10.3. As discussed in Chap. 10, such bolt-hole tears are likely to occur when the bolts subjected to the higher prying action are located on the flexural tension crest of the corrugations.

It is interesting to note that, as discussed in Chap. 10, the desirability or undesirability of joints with two bolts per pitch of the corrugations does not change with the direction of the moments applied to the joint. In the case of joints with four bolts per pitch of the corrugations, also, the direction of applied moment is irrelevant with respect to desirability of the bolting arrangement. From consideration of the formation of bolt-hole tears, this joint is undesirable because it always contains bolts in those critical locations which are prone to the formation of bolt-hole tears.

A joint with three bolts per pitch of corrugations can be formed by adding the extra bolts to the basic arrangement of two bolts per pitch. If the basic arrangement is undesirable, then the resulting joint with three bolts per pitch is also undesirable regardless of where the extra bolts are added; an example of such a joint is shown in Fig. B.1d. The question of whether the joint with three bolts per pitch formed by

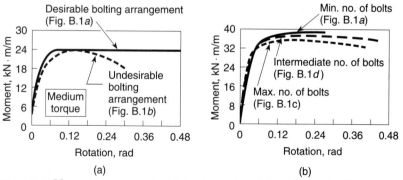

Figure B.3 Moment-rotation plots obtained experimentally: (a) plots for 5-mm-thick plates; (b) plots for 7-mm-thick plates with medium torque on bolts.

adding extra bolts to the desirable basic arrangement of Fig. B.1a, will be desirable or not depends, however, upon the direction of applied moments. For example, consider the joint shown in Fig. B.1e, which is subjected to positive moments causing tension at the bottom face; in this case, half the bolts with higher prying force, i.e., those closer to the hidden edge, are located on the flexural tension side. As a result of this, the joint is deemed to be undesirable. The situation, however, reverses when this same joint is subjected to negative moments as shown in Fig. B.1f. It can be readily appreciated that this arrangement can now be considered desirable.

Moment-rotation plots of joints with two, three, and four bolts per corrugation pitch are given in Fig. B.3b for 7-mm- (0.276 in) thick plates and for bolts with medium torques. It can be seen in this figure that the increase in the number of bolts per pitch, in fact, decreases the flexural strength of the joint. From considerations of moment capacity alone, the joint with the minimum number of bolts, i.e., with bolting arrangement 1, is the preferred one. The moment rotation plots for this preferred joint are given in Fig. B.4 for plates with different thicknesses.

B.3.2 Effect of bolt torque

The joints for the experimental study (Refs. 1 and 2) were tested with 20-mm- (¾ in) diameter bolts torqued to 500, 250, and 50 N·m (369, 184, and 37 lb·ft); these three levels are referred herein as high, medium, and low.

It was found that the effect of bolt torque on the load-deformation characteristics of the joint is not influenced by the arrangement of the bolts. An increase in the bolt torque was found to enhance both the

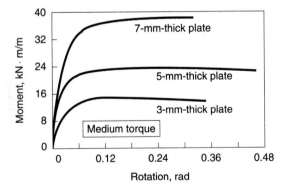

Figure B.4 Moment-rotation plots obtained experimentally for corrugated plates with the bolting arrangement of Fig. B.1 (a).

stiffness and the strength of the joint, as seen in Fig. B.5, which shows the moment-rotation plots of joints with 5-mm- (0.197 in) thick plates bolted according to the scheme of Fig. B.1a.

From the results of the aforementioned tests, it seems obvious that the highest torque is the most desirable one. However, it is recalled that these tests were conducted on straight corrugated plates, whereas in practice the plates are curved and the overlapping segments may not always have the same curvature. To avoid the creation of stress-raiser by forcing plates of different radii of curvature together, high torques should not be used to try to improve poor nesting of plates. Medium torques with good nesting are preferable. It is noted that the *Ontario Highway Bridge Design Code* (Ref. 3) requires that the bolt torques should lie between 200 and 340 N·m (147 and 251 lb·ft.)

B.3.3 Effect of washers

The formation of bolt-hole tears in undesirable bolting arrangements is believed to be initiated from stress-raisers caused by the bolt heads or nuts indenting the corrugated plates. One possible method of eliminating such indentations is by the provision of washers between the plate and the bolt head or nut. On the basis of a limited number of tests with and without washers, it is concluded that even shaped washers are not effective in eliminating the stress-raisers (Refs. 1 and 2).

B.4 Recommendations

On the basis of the preceding work, the following recommendations are made with respect to the bolted connections of corrugated plates

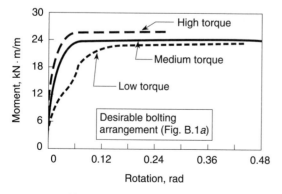

Figure B.5 Moment-rotation plots for 5-mm-thick plates illustrating the effect of bolt torque.

employed in soil-steel bridges. Because of their higher ductility, only those joints are recommended which are labeled as desirable in Fig. B.1. As can be seen in this figure, the joints with four bolts per corrugation pitch are not considered desirable and should not be used. It is possible to use a desirable arrangement of bolts in joints with three bolts per corrugation pitch; however, the achievement of such arrangement in practice is not without the possibility of inadvertent errors in the field.

Even shaped washers do not seem to eliminate stress-raisers in the corrugated plate. The use of washers is, therefore, not recommended.

References

1. Lee, R. W. S., and Kennedy, D. J. L., *Behaviour of Bolted Joints of Corrugated Steel Plates,* Structural Engineering Report 155, Department of Civil Engineering, University of Alberta, Edmonton, Alberta, Canada. 1988
2. Mikhailovsky, L., Kennedy, D. J. L., and Lee, R. W. S., "Flexural Behaviour of Bolted Joints of Corrugated Steel Plates," *Canadian Journal of Civil Engineering,* vol. 19, no. 5, 1992, pp. 896–905.
3. *Ontario Highway Bridge Design Code,* Ministry of Transportation of Ontario, Downsview, Ontario, Canada, 1992.

Index

AASHTO (American Association of State Highway and Transportation Officials):
  bridge design specifications, 74, 86, 108, 118, 119, 129, 131, 173, 174, 204, 219, 220, 234
  buckling stress, 162
  crushing stress, 162
AASHTO, Drainage Manual, 288
AASHTO, Method, 158, 170
AASHTO, Technical Guidelines, 288
ABC (Arch-Beam-Culvert), 254
Abdel-Sayed, G., 71, 103, 114, 122, 124, 131, 132, 135, 203, 227, 234, 237, 251, 253, 271
Afflux, 274
Agarwal, A. C., 48, 69, 261, 262, 271, 293, 294, 335
Aggressive ions, 37
AISI (American Iron and Steel Institute), 41, 81, 85, 94, 118, 119, 131, 219, 234, 288
Akl, A.Y., 240, 241, 242, 245, 271
Alberta DOT, 288
Allgood, C. R., 94, 99, 101, 132
American Concrete Pavement Association, 263
Angle of internal friction, 218
Anode, 31
Anodic reaction, 31
Arching, 18, 48
  negative, 49
  positive, 48
  transverse, 58
Arching factor, 94
ASP (American Society for Photogrammetry), 319
ASTM (American Society for Testing and Materials), 204, 214, 234

Atterberg's consistency limit, 206
Axial rigidity parameter, 102, 107, 137

Backfill, 8, 163
Backfill, extent of, 150, 194
Backfill placement, 196
Backfill placement over the top, 198
Backfill placement under the haunches, 196
Backfill selection requirements, 218
Backfilling, 194
Baikie, L. D., 114, 115, 116, 132, 230, 232, 235
Bakht, B., 1, 29, 43, 48, 69, 103, 124, 131, 135, 175, 237, 248, 261, 262, 271, 291, 293, 294, 335
Bearing failure, 297
Beaton, J. L., 41
Bedding, 9, 182
Belke, T., 344
Bending stiffness parameter, 107
Bevelled end, 19
Bishop, A.W., 225, 234
Bolt torque, 348
Bolt-hole tears, 292, 293
Bolt holes, slotted, 239
Bolted connections, 243
Bolted joints, failure loads of, 344
Bolting arrangement, 189
Bolting arrangement 1, 16, 344
Bolting arrangement 2, 16, 344
Bolting arrangement 3, 16, 344
Booy, C., 114, 131
Boully, G. K., 255, 256, 271
Bowles, J. E., 235
Brewer, W. E., 263, 264, 265, 269, 271
Bridge, 6
  (*See also* AASHTO: bridge design specifications)

ABOUT THE AUTHORS

GEORGE ABDEL-SAYED is a professor in the department of civil and environmental engineering at the University of Windsor, and chairman of the Technical Committee on Soil-Steel Structures of the Canadian Society for Civil Engineering (CSCE).

BAIDAR BAKHT is principal research engineer for the Ministry of Transportation of Ontario, and vice-chairman of the Technical Committee for the Canadian Highway Bridge Code.

LESLIE G. JAEGER is emeritus research professor of civil engineering and applied mathematics at the Technical University of Nova Scotia, and an associate of Vaughan Engineering Associates Ltd.